Intelligent Processing Algorithms and Applications for GPS Positioning Data of Qinghai-Tibet Railway

Dewang Chen · Ruijun Cheng

Intelligent Processing Algorithms and Applications for GPS Positioning Data of Qinghai-Tibet Railway

Springer

Dewang Chen
Fuzhou University
Fuzhou, China

Ruijun Cheng
SKL of Rail Traffic Control and Safety
Beijing Jiaotong University
Beijing, China

ISBN 978-3-662-58968-7 ISBN 978-3-662-58970-0 (eBook)
https://doi.org/10.1007/978-3-662-58970-0

Jointly published with Beijing Jiaotong University Press, Beijing, China
The printed edition is not for sale in Mainland of China. Customers from Mainland of China please order the print book from: Beijing Jiaotong University Press.

This Springer imprint is published by the registered company Springer-Verlag GmbH, DE part of Springer Nature.
The registered company address is: Heidelberger Platz 3, 14197 Berlin, Germany

Contents

Chapter 1
Application of Satellite Positioning in Railway

1.1 Overview

Satellite navigation system is a satellite system that covers global space to realize positioning. It determines its position through electronic receiver, namely longitude, latitude, and altitude, and ensures that the time signal transmitted through satellite broadcasting is within the error range of 10 m. Each satellite navigation system covering the globe is usually composed of a constellation of 20–30 satellites, spreading over several orbital planes. In the actual system, the number of satellites in each satellite navigation system is not identical, but the satellite orbital tilt is greater than 50°, the orbital period is about 12 h, the height is about 20,000 km distance from the surface of the earth [1]. Figure 1.1 shows a schematic of a satellite in orbit. The satellite navigation system has the characteristics of all-weather positioning, global positioning, three-dimensional high-precision positioning, and dynamic navigation, so it has a wide range of applications.

Countries around the world, including the USA, Russia, the European Union, and China, are also working on their own navigation systems to avoid being held hostage. At present, only American global positioning system (GPS) and the former GLONASS system of Soviet Union are the navigation system that covers the globe. China Beidou satellite (BDS) navigation system has been in service in Asia Pacific at the end of 2012 and is expected to realize global coverage in 2020. European Union's Galileo positioning system is scheduled to be fully operational in 2020 at the earliest. France, Japan, and India are building regional navigation systems.

D. Chen and R. Cheng, *Intelligent Processing Algorithms and Applications for GPS Positioning Data of Qinghai-Tibet Railway*, https://doi.org/10.1007/978-3-662-58970-0_1

Fig. 1.1 A schematic of a satellite in orbit

1.2 Satellite Navigation

The world's four major satellite navigation systems are made up of the US GPS global positioning system, Russia's GLONASS, Chinese BDS Beidou satellite navigation system, and the European Union's Galileo navigation system.

1.2.1 Global Positioning System

Global positioning system (GPS) is a new-generation space satellite navigation and positioning system of medium-range circular orbit developed and maintained by the department of defense of USA in the 1970s (Fig. 1.2). Its main purpose is to provide real-time, all-weather, and global navigation services for three fields of sea, land, and air. After more than 20 years of research and experiments, at a cost of 30 billion dollars, 24 GPS satellite constellations covering 98% of the world had been set up by 1994. It can provide accurate positioning, speed measurement, and high-precision time standard for most parts of the earth's surface. Users only need to have a GPS receiver to obtain the service without extra payment. The implementation of GPS plan has gone through three stages [2, 3]:

The first stage is the stage of program demonstration and preliminary design. From 1978 to 1979, four experimental satellites were launched by a Gemini rocket

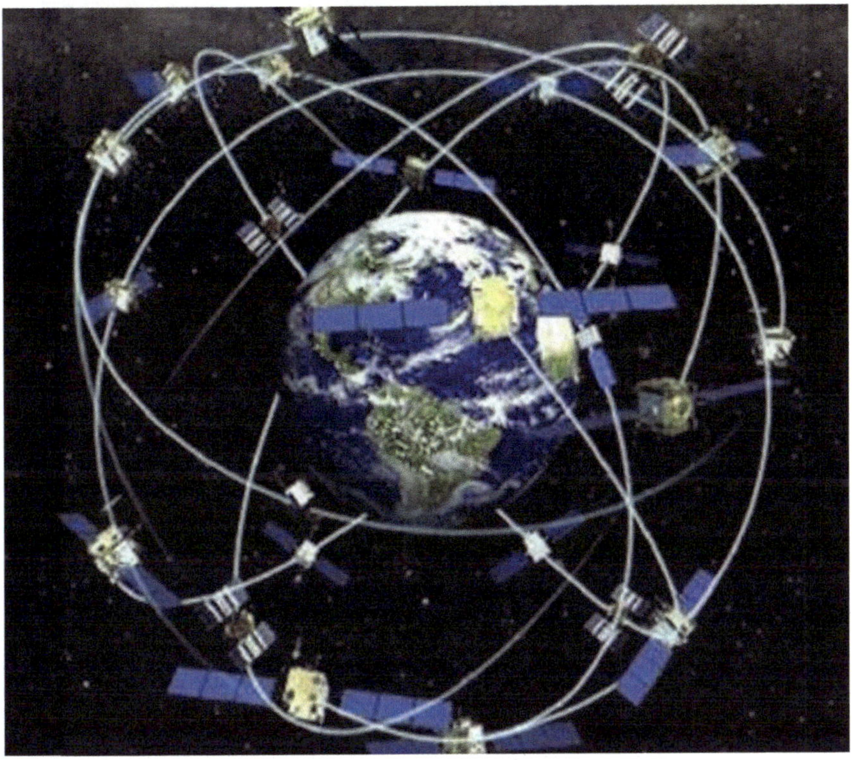

Fig. 1.2 GPS satellite working constellation diagram

from Vandenberg air force base in California. The orbit of the satellites is 26,560 km with an inclination of 64°. The orbital altitude is 20,000 km. In this stage, a ground receiver is mainly developed and a ground tracking network is set up.

The second stage is the comprehensive development and testing stage, which covers the period from 1979 to 1984. Following the first phase, seven experimental satellites, called BLOCK I, were launched and various receivers were developed.

The third stage is the practical networking stage from 1989 to 1994. This phase of the satellite launch, called BLOCK II and BLOCK IIA, declared the GPS system is applied into engineering.

The GPS global positioning system consists of three parts: the space satellite part, the ground monitoring part, and the user equipment part. The components of each part are as follows.

(1) Space satellite part. The GPS satellite constellation consists of 24 satellites, evenly distributed on six orbital planes, that is, four satellites are on each orbital plane, among which 21 are working satellites and the other three are standby satellites. The angle between the satellite orbit plane and the surface of the

earth's equator orbit is 55°, the right ascension of the ascending node for every orbital plane is 60°, and the argument of latitude of the satellite orbit on an orbital plane has an advance angle of 30° than that of the corresponding satellite of its west adjacent plane. The purpose of this layout is to ensure that at least four satellites can be observed at any place and at any time in the world, to meet the need of precise navigation and positioning.

(2) Ground monitoring. The ground monitoring part mainly consists of one Master Control Station (MCS), four Ground Antenna Stations (GAS), and six Monitor Stations (MS). The monitoring station transmits the satellite observation data to the MCS after preliminary processing. The MCS collects tracking data from each monitoring station, calculates the satellite's orbit and clock parameters, and sends the results to the GAS. The GAS injects the collected data and the command of the MCS into the satellite when each satellite is operating above it. The MCS is the management center and technical center of the whole ground monitoring system.

(3) User equipment section. The user equipment is mainly GPS receiver. The main function is to calculate the user's three-dimensional position and time by receiving the signal from the GPS satellite and using the information transmitted. This part mainly consists of GPS receiver hardware and corresponding data processing software. The hardware of GPS receiver includes the host, antenna, and power supply of the receiver. Its main function is to receive the signals sent by the satellite so as to obtain necessary navigation and positioning information and realize navigation and positioning through simple data processing.

The main functions of GPS are to provide accurate timing for the base station of the observatory and the communication system, to provide precise positioning function for vehicle anti-theft system, mobile phone, electronic map, and positioning system, and to play an important role in the navigation of weapons, vehicles, ships, and aircraft. GPS is also widely used in various engineering surveys, such as the construction of roads, bridges, and tunnels by using GPS equipment [4].

1.2.2 GLONASS

GLONASS is the Russian acronym for global satellite navigation system. Established by the Soviet Union in 1976, the GLONASS positioning system is now operated by the Russian government and, similar to GPS, has a civilian window. GLONASS was established in 1991 with a global satellite navigation system. By 2012, the number of satellites in the system had increased to 30 for global positioning navigation, and its satellite navigation range could cover the entire earth surface and near-earth space with positioning accuracy of about 1 m. A well-deployed navigation and positioning system will be widely used in defense construction, aerospace, transportation, geological exploration, disaster relief, and other fields [2, 3].

Similar to GPS created by USA, GLONASS system comprises of three parts: satellite, ground control stations, and user equipment. The current system includes 21 work satellites and three backup satellites. Unlike GPS, these 24 satellites only distributed in three orbital planes. So, each orbit plane has eight satellite, the orbit altitude is 19,000 km, and the running cycle is 11.25 h. In terms of technology, GLONASS differs from GPS in the following three aspects:

(1) The transmitted frequency of satellite is different. GPS satellite signals adopt code division multiple access (CDMA) system. So, each satellite's signal frequency and modulation mode are the same, and the signals of different satellites are determined by different pseudo-code. However, since GLONASS adopts frequency division multiple access (FDMA) mechanism, satellites are differentiated by different frequencies. Therefore, the GLONASS positioning system can prevent the entire satellite navigation system from being interfered by the enemy and has stronger anti-interference ability.
(2) The coordinate system is different. GPS uses the world geodetic coordinate system (wgs-84), while GLONASS uses the former Soviet geocentric coordinate system (pz-90).
(3) Different standards of time. GPS system time is associated with world coordinated time, while GLONASS is associated with Moscow standard time.

The emergence of the GLONASS system broke the exclusive business of the USA in the field of satellite navigation system. This allows the user to choose when using, or combine the two to improve positioning accuracy. At present, GLONASS system is mainly used for navigation and positioning and can also be used for various positioning, measurement, etc.

1.2.3 Beidou Satellite Navigation System (BDS)

Beidou satellite navigation system is a global satellite navigation system that is being implemented and independently operated in China. It is committed to providing high-quality positioning, navigation, and timing services to global users and can provide further services to authorized users with higher requirements. Its construction and development follow the four principles of openness, autonomy, compatibility, and gradual progress. The diagram of Beidou satellite navigation system is shown in Fig. 1.3. The construction of the system includes the experimental system and the formal system [5].

The test system, also known as "Beidou No.1," was officially developed in 1994 and officially approved in 1994. The three navigation satellites were successfully launched on October 31, 2000, December 21, 2000, and May 25, 2003, the last of

Fig. 1.3 Schematic diagram of Beidou satellite navigation system

which was a spare satellite. The service scope of the test system is from 70° to 140° for east longitude, from 5° to 55° for north latitude. The three-dimensional positioning accuracy is dozens of meters, and the horizontal positioning accuracy is 100 m. In addition, the positioning accuracy can be improved to about 20 m after the check station is established. The timing accuracy is 20 ns, and the positioning response time is 1 s. The capacity is 540,000 users per hour. During the 2008 Olympic Games, Beidou No.1 and GPS satellite positioning system played the role of double insurance in traffic and venues' safety monitoring. The test system consists of three parts, space segment, ground segment, and user segment:

(1) Space segment is composed of three geostationary satellites. Two work satellite positioning is used from 80° and 140° for east longitude above the equator. The third satellite is located in east longitude 110.5° and is taken as backup satellite, which can work if one of the first two working satellites fails.
(2) Ground segment consists of a central control system and a calibration system. The central control system is mainly used for determining satellite orbit, ionospheric correction, user position determination, user short message exchange, etc. Calibration system can provide distance measurement and calibration parameters.
(3) User segment is the user's terminal. According to the different application environment and function of user machine, user terminal contains five types, which are ordinary type, communication type, timing type, command type, and multi-model user machine.

The test system uses a small number of satellites to realize active positioning, so the cost is relatively low, but the system is limited in positioning accuracy, user capacity, and other performance. Therefore, China officially launched the construction of Beidou No. 2 satellite navigation system with global navigation capacity in 2004, namely the formal system of Beidou satellite navigation system, which is a dual-use system for both military and civil purposes. Since 2012, the system has been providing formal services to most of the Asia-Pacific region, covering south latitude 55° to north latitude 55°, east longitude 55° to east longitude 180°. The construction of the global system will be completed by 2020. The navigation system provides two service modes, open service and authorization service. The open service provides free, stable, and reliable basic positioning, speed measurement, and timing services for global users, with positioning accuracy of 10 m, speed measurement accuracy of 0.2 m/s, and timing accuracy of 20 ns. The authorized service provides global users with more accurate positioning, navigation, speed measurement, and timing services.

Like the experimental system, the formal system consists of three parts: space segment, ground segment, and user segment. The space segment program consists of 35 satellites, including five geostationary orbit satellites, 27 medium-circle orbit satellites, and three slant synchronous orbit satellites. Five geostationary orbit satellites were located in east longitude 58.75°, 80°, and 110.5°, 140°, and 160°, respectively. The ground segment consists of several ground stations such as the main control station, injection station, and monitoring station. The user segment includes Beidou

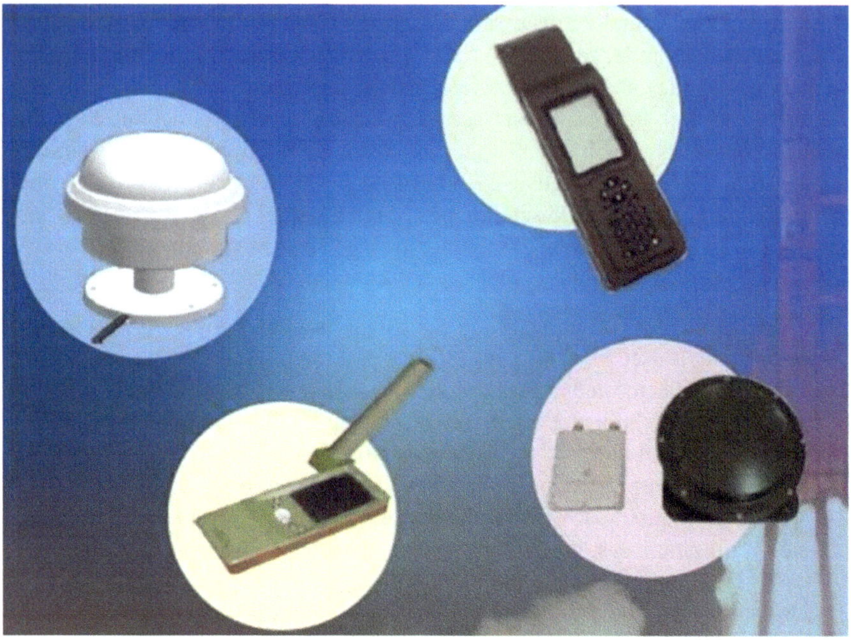

Fig. 1.4 Various terminal devices based on GPS

user terminal and terminal compatible with other satellite navigation systems [6, 7]. Various types of terminals are shown in Fig. 1.4.

1.2.4 Galileo Global Positioning System

Galileo global positioning system is a satellite positioning system under construction in the European Union. It is the fourth positioning system that can provide the civil service after GPS of the US GLONASS of Russia and Beidou satellite navigation system of China. Galileo is designed to provide users with more accurate data, while increasing coverage at higher latitudes, such as places like Norway and Sweden, and reduce reliance on the US global positioning system, especially during wars. The development of the system mainly goes through the following four stages [4]:

From 1999 to 2000, the system feasibility was assessed and defined in the first stage. From 2001 to 2005, the system was developed and tested in the second stage. The main work during this stage includes gathering mission requirements, developing 2–4 satellites, ground parts, and in-orbit verification of the system. The third stage, from 2006 to 2007, is the deployment stage of the satellite. The main tasks during this stage include developing the satellite launch network, the construction of ground stations, and the overall coordination of the system. The fourth stage started in 2008, when the system was put into use.

Galileo will launch a total of 30 satellites, 27 of which will serve as working satellites and the remaining three as backup satellites. The 30 satellites are located in three orbital planes with an inclination of 56°. Each orbit has 10 satellites, nine of which operate and the last one serves as a backup satellite. Each satellite has a height of 23,222 km above the ground, a life span of over 12 years, a weight of 675 kg, a length, width, and height of 2.7 m*1.2 m*1.1 m, a solar collector plate width of 18.7 m and a power of 1500w. Galileo's service is divided into two parts: free service and paid service, including the following 6 types:

(1) Open service mainly provides users with free positioning, navigation, standard time, etc. The service target is the general public, such as the navigation system of general automobiles.
(2) Commercial service mainly refers to additional services not included in the public service, such as adding encrypted data in the public service.
(3) Safety-of-life service. This part of service is only applicable to traffic transportation, guiding vessels to enter port, railway transportation, traffic control, etc.
(4) Public Regulated Services are available only to EU member states.
(5) Navigation service provided by local components for regional organizations provides more accurate positioning and timing services to users with such requirements.
(6) The service for search and rescue system, Galileo satellite navigation system's rescue function is basically the same as the previous international general satellite search and rescue system (COSPAS-SARSAT), but the performance has been greatly improved.

1.3 Foreign Application

The first application of GPS positioning system to railway trains is ARES, developed by American Burlington northern in 1984 in partnership with Rock Well. At present, European railways are strengthening the use of GPS technology and setting up differential base stations along the railway, so as to combine it with mobile communication technology to improve railway transport capacity. The German railway ministry believes that the GPS satellite navigation system can replace the existing routine operation, and the satellite navigation, mobile communication, and guidance system can replace the existing fixed signal system on the test line. Because GPS train positioning does not depend on the railway's existing fixed signal system to locate the train, it can guide the driver or the automatic control driving system to drive the train in a safe and economic way. Railway projects adopting GPS technology include incremental train control system (ITCS), ATLAS400, GADEROS, LOCOPROL/LOCOLOC, RUNE, etc. GPS-based train positioning technology will be an important development direction of train operation control technology in the future.

(1) Train Control System Based on LOCOPROL

Low-cost satellite-based train location system for signaling and train protection for low-density traffic railway lines (LOCOPROL) is a train operation control system developed in Europe based on satellite positioning. This is a low-volume railway line signal system with low-cost characteristics. The biggest innovation of the system is the introduction of global satellite positioning technology for the railway signal system. The system did not adopt a new design concept, but adopted the concept of an existing system, thus reducing the system design risk and development difficulty [8, 9]. The specification of ETCS system is an important basis for the automatic control system of LOCOPROL system.

Using GPS, GLONASS, Galileo, and other global satellite positioning systems, LOCOPROL has developed a low-cost satellite positioning system with high fault-safety. Different from the traditional method, the LOCOPROL system uses the data of several satellites to calculate the confidence interval. After processing the information of the confidence interval, the fault-safety location data can be obtained. The basic principle of the system is shown in Fig. 1.5.

In the aspect of system composition, LOCOPROL system adopts distributed architecture, which improves the ability of system extension. One of the important designs is to reduce the input cost and maintenance cost of the system by reducing the number of trackside equipment as much as possible. In the traditional signal system, a large number of train occupancy detection devices, train positioning, signal machine, beacon, and so on are required to be set up beside the rail track or on the train. In this system, the number of trackside equipment is reduced to the minimum through the application of locomotive signals for replacing the train occupancy detection device and the use of ground signs for replacing the beacon position marking of the signal machine and switch machine.

In terms of accuracy, the train position obtained by the system algorithm is only for a certain line zone, rather than a certain point where the train is, and the actual position of the train should be at one end of the calculated line segment. The accuracy of train positioning depends on the probability of measurement. In terms of train integrity check, 3 pairs of satellite from different sources are used to ensure the required integrity indicators. The system can provide the absolute position of the train on the line under the premise of fault-safety. At the same time, if a beacon is set on the line, the satellite positioning of train data can be combined with the location data provided by the beacon to locate the exact position of the train.

Compared with the traditional positioning technology, the sensor used in LOCOPROL satellite positioning does not need regular calibration. Since the equipment is installed on the train and is highly adaptable to the environment, the weather conditions such as rain, snow, fog, and the type of rail will not affect the system information. The LOCOPROL system has passed strict safety tests and evaluated and verified the following aspects from the perspectives of severity, probability of

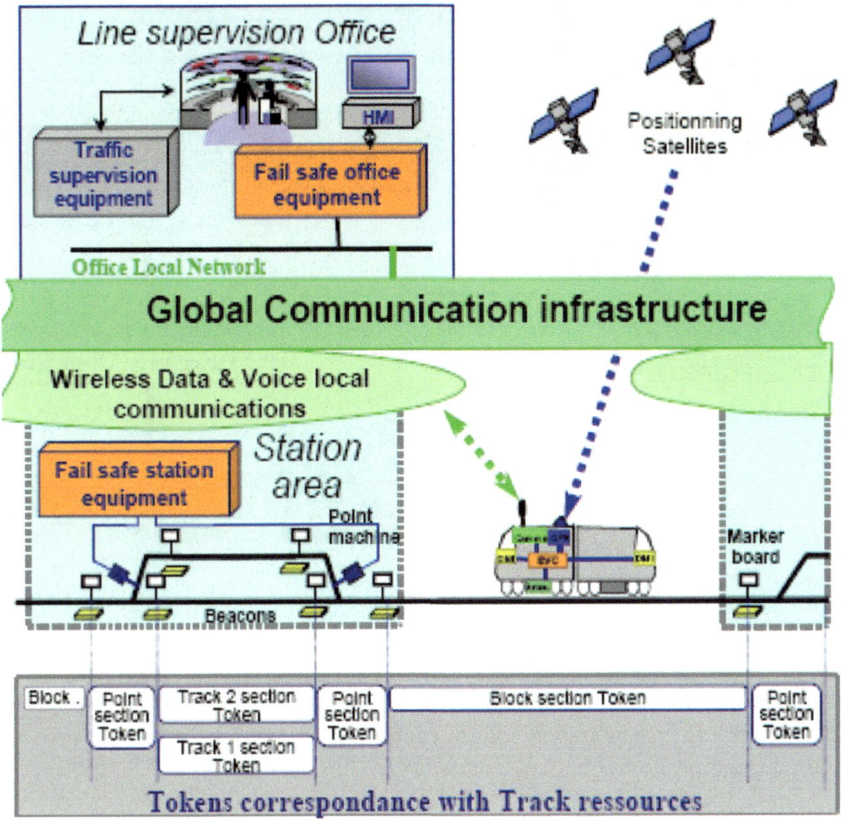

Fig. 1.5 Basic principle of the system

occurrence, and solution: global positioning satellite, satellite ground equipment, space environment, ground environment, noise, and data blockage.

(2) Train Control System Based on GADEROS

Under the framework of the European railway transport management system (ERTMS) and European train operation control system (ETCS), Galileo Demonstrator for Railway Operation System (GADEROS) uses the satellite technology to realize train positioning and is widely used in conventional and low-density railway systems. It is funded by the European Unit, and several European countries, including Britain, France, and Spain, are jointly involved in the research and development of the project [10, 11].

The research content of this project includes the following aspects: (1) analysis of railway demand for train positioning; (2) determine the functions and interfaces of GNSS positioning system under the framework of ERTMS/ETCS; (3) put forward test base and test plan of demonstration project; (4) necessary test equipment

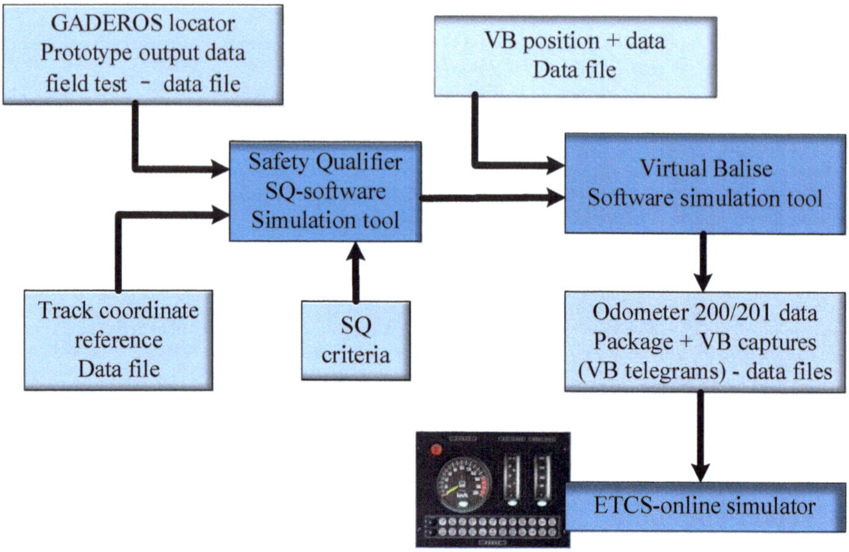

Fig. 1.6 Composition of virtual balise in GADEROS system

required in the development process; (5) simulate GNSS locator in ERTMS/ETSC laboratory; (6) integrate various protocols and conduct field tests; (7) application of GNSS locator in transportation control center; (8) evaluate the project. Figure 1.6 shows the composition of virtual balises in GADEROS system.

1.4 Domestic Application

In China, GPS positioning technology was first applied to train positioning in the Golmud–Lhasa section of Qinghai–Tibet railway. In addition, the railway mobile communication system (GSM-R) was used as the platform for vehicle-ground information transmission to realize the signal system, which integrates interlocking, blocking, and speed protection of trains. The signal system adopts the incremental train control system (ITCS), which combines the wheel sensor with the differential GPS positioning system to achieve accurate train positioning. The system can not only shorten the departure time and improve the passing capacity of the train, but also reduce the equipment on the side of the rail. The functional schematic diagram of the signal system is shown in Fig. 1.7. The functions of the ITCS system mainly include the following three types [12–14]:

(1) Virtual Automatic Block and Station Block

ITCS system divides the interval between stations of the whole line into several virtual block partitions, and the system sets up virtual signal machines to protect

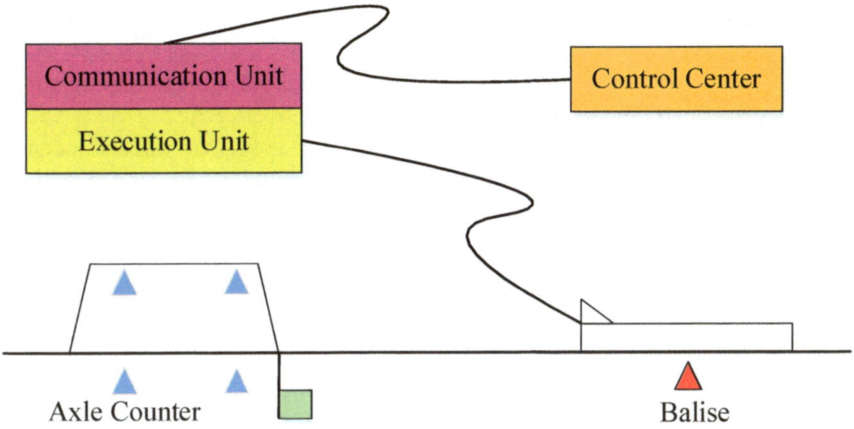

Fig. 1.7 Function schematic diagram of the signal system

each virtual block partition. The signal machine adopts two kinds of control signals, such as forbidden signals and permitted signals.

(2) Realize Interlock Control of Unattended Station

The interlock of unattended station is realized by VHLC. Under the principle of "failure-safety," the interlock operation is realized by receiving and executing the dispatching command of the dispatching center, and thus the selection and opening of the train route are completed.

(3) Train Operation Control and Speed Protection

ITCS system uses GSM-R wireless communication network to transmit the messaging between ground equipment and onboard equipment and obtain the status information of virtual signals from RBC on the ground and VHLC. ITCS mainly applied onboard signals to control train operation and follows the principle of human-control priority. The speed protection system only works when the train runs faster than the permitted speed. Its fundamental purpose is to guarantee the safety of the train.

The train operation control system of QTR line has realized the integration of signal control, computer technology, and modern communication. In terms of performance, the system realizes less maintenance, low energy consumption, and unattended station. The system consists of the following five subsystems [15, 16]:

(1) Special communication system for highland area. Due to the harsh environment of QTR line, GSM-R wireless communication network cannot be adopted, and more reliable network should be applied to guarantee the high reliability of information transmission. The system includes high reliability information network of both control center and station and the wireless data transmission network within station range.

(2) Regional computer-based interlocking system. The system includes centralized interlocking remote control system, automatic inter-station block control, traffic dispatch system, and wireless locomotive signal control system. The computers that make up these systems are all safety computers.
(3) The wireless locomotive signal system is composed of arrival wireless locomotive signal, highly reliable subjectivity, safety closed loop inspection vehicle, and intelligent onboard equipment.
(4) The wayside equipment system with low energy consumption is composed of the low energy storage turnout switch system and the redundant system of multiple operating modes.
(5) Maintenance-free integrated power supply system includes integrated power supply and automatic conversion centralized monitoring.

There are five operating modes of ITCS signal system, including starting test mode, ITCS entry mode, station mode, ITCS departure mode, and tail-end mode. The entering conditions of various working modes for trains are described as follows [17, 18]:

(1) Starting test mode. The main content of this working mode is to confirm that the onboard computer and peripheral equipment are in normal working state.
(2) ITCS entry mode. At this time, ITCS is in the working state to control the train. Under this mode, the onboard computer displays the minimum speed limit, line speed, temporary speed limit command, or other speed limit for special conditions.
(3) Station mode. When the switch is in the "entry" position and the conditions of ITCS entry mode are not met, the onboard computer works in the station mode. When the train runs in station mode, the speed of the train will always be limited to a certain value, and the speed limit value is determined by the type of running train.
(4) Departure mode. When the switch is in the "leave" position and the ITCS entry mode is not satisfied, the onboard computer is working in the leave mode. At this time, the onboard computer does not provide overspeed protection.
(5) Tail-end mode, when the onboard computer works in the tail-end mode, ITCS does not control the train.

In addition to the QTR line, the GPS positioning system is also widely used in various engineering surveys, such as the railway line survey. In the course of railway laying, it is used to guarantee the safety of railway laying locomotive. It is also used in railway logistics to track freight trains, improve the service quality of railway transportation, and make operation more transparent. Furthermore, it is also used in the railway dispatch system. The details are described as follows:

(1) Railway Survey

The static and dynamic functions of GPS are mainly included in the practical railway measurement. The static function is to determine the three-dimensional coordinates of a certain point on the ground through the received satellite information [7]. The dynamic function is to place the given three-dimensional coordinate point position on the actual ground through the satellite navigation and positioning system.

(2) Railway Laying Operations

In the railway laying operation, the locomotive has no safety equipment except a wireless telephone. Because the normal system of train operation is not established, the safety of train operation is not guaranteed. The positioning function of satellite navigation system can ensure the orderly and controllable track-laying operation and the safety of track-laying operation train.

(3) Rescue Train

The satellite navigation system is applied to locate and monitor the accident train and rescue train in real time.

(4) Application in Scheduling System

The railway dispatching system needs to monitor the train location in real time to ensure the safety of train operation. Through GPS technology and the existing wireless train scheduling system, train signal, monitoring device and train number identification equipment, the navigation and positioning system can in real time to send the train operation time, train number, train position, and speed information to the scheduling monitoring center. The computer can display the various states information of the train on the big screen through GIS, which makes the station attendant and dispatcher directly understand and grasp the location information of the train.

1.5 Brief Summary

This chapter mainly introduces the four relatively complete global navigation and positioning systems in the world, including GPS, GLONASS positioning system, Beidou satellite positioning system, and Galileo global navigation system. These four positioning systems have their own advantages and disadvantages. GPS has the advantage of global coverage, unlimited number of users, and the ability to provide continuous, real-time positioning services. It only takes a few seconds to more than ten seconds for a single positioning time. The user does not need to send any electromagnetic signal, as long as he receives the navigation signal and positioning signal of the satellite. The GLONASS system has been modernized, which has stronger positioning accuracy. The GALILEO satellite adopts medium earth orbit, of which the typical function is signal relay, that is, data transmission can be transmitted to user receiver through mobile communication network. The receiver can not only receive the signals of the system, but also receive the signals of GPS and GALILEO, and combine the navigation function with mobile phone function, which is a great advantage of the GALILEO system. The Beidou satellite navigation system developed in China has the short message communication function, which the GPS system does not have. The "Beidou" system is the active bidirectional ranging two-dimensional navigation, which is that the ground central control system provides users with three-dimensional positioning data by calculation. The Beidou satellite system is characterized by fewer satellites, less investment, simpler and cheaper user equipment. Among the four navigation and positioning systems, the development of GPS is the most mature, so it is also the first widely used.

This chapter also introduces the application of GPS in railway, mainly the application of GPS in Chinese QTR line, including the function, composition, and operation mode of the system.

References

1. Han NY. GPS satellite navigation system. J Shenzhen Univ (Sci & Eng). 1989;6(1):105–12.
2. Luo X, Cai C, Dai W, Kuang C. Contrastive analysis of navigation signals for Galileo satellites and GPS satellites. In: Proceedings of the fourth China satellite navigation academic annual conference–S2 satellite navigation signal system and its compatibility and interoperability. 2013.
3. Ma L, Kong Y, Guo SH, Song HY, Pang Y. Comparison of several satellite navigation systems. Mod Surveying Mapp. 2011;34(3):3–6.
4. Huang AM. The Galileo project impact to American GPS. Geomatics Spat Inf Technol. 2007;30(3):41–4.
5. Qiu ZJ. Introduction to BEIDOU satellite navigation system and other satellite systems. Sci Technol Inf. 2012;36:27.
6. Yang YX. Progress, contribution and challenges of compass/BEIDOU satellite navigation system. Acta Geodaet Cartographica Sin. 2010;39(1):1–6.
7. Zhou L, Liu BZ. Technical feature analysis and application of BEIDOU satellite positioning system. GNSS World China. 2004;29(4):12–6.
8. Mertens P, Franckart JP, Starck A. LOCOPROL: a low cost train location and signalling system for low density. Lines. 2003. Retrieved from 11 Oct 2008. http://www.ertico.com/download/locoprol%20documents/wcrr03v3.pdf.
9. Mertens P, Franckart JP, Starck A. A low-cost train location and signalling system for the low-density lines. Edinburgh: WCRR; 2003.
10. Urech A, Pérez Diestro J, Gonzalez O. GADEROS, a GALILEO demonstrator for railway operation system. Data Syst Aerosp. 2002;509.
11. Barbu G. GALILEO Demonstrator for Railway Operation System Final Report. INECO, http://galileo.uic.asso.fr/docs/geTd1003_final_report.pdf, 18 Jan 2005.
12. Tang JG. Research and application of comprehensive monitoring technology for train operation control system of Qinghai-Tibet railway. CHINESE Railways. 2008;1:55–8.
13. Zhang XZ. Development on project of interlocking based on area computer control in Qing-Zang railway. Railway Comput Appl. 2008;12(4):39–42.
14. Liang H. Analysis on the function and dynamic debugging of ITCS system of Qing-Zang line. Electr Railway. 2009;3:48–50.
15. Chen Z, Y, Cao Y, Zhang DL, Yang L. Study on virtual occlusion in ITCS system of Qinghai-Tibet railway. Railway Signal Commun. 2007; 43(12):3–5.
16. Huang M, Wang CL, Cao JP. Application of ITCS based on GSM-R to Qingzang railway. Railway Comput Appl. 2006;15(7):15–8.

17. Liu W, Wang XM. First check of ITCS in Qinghai-Tibet railway. Comput Commun Signal. 2009;18(9):52–4.
18. Jin J, Wang CL. Research on ITCS communication and signal system of Qingzang railway. Inf Commun. 2008;21(4):60–3.

Chapter 2
Qinghai–Tibet Railway

2.1 Overview

The Qinghai–Tibet plateau, with an average altitude of more than 4000 m, is known as the "roof of the world" and the "third pole of the earth" and has long been regarded as a "forbidden zone of human life." Qinghai–Tibet railway (QTR) is a landmark project to implement the Western development strategy and one of the four major projects in the new century in China. The railway with the length of 1956 km starts from Xining, Qinghai in the east and ends at Lhasa, Xizhang in the west. Among them, the 814 km section from Xining to Golmud was completed in 1979 and put into operation in 1984. The QTR starts from Golmud, passes through Nachitai, Wudaoliang, Tuotuohe, Yanshiping, crosses Tanggula mountain, goes through Ando, Naqu, Dangxiong, Yangbajing of Tibet autonomous region, and ends at Lhasa, with a total length of 1142 km. Among them, the newly built line is 1110 km, which was officially started on June 29, 2001, and officially opened to traffic on July 1, 2006 [1–3]. Figure 2.1 shows the schematic diagram of the line between Golmud and Lhasa of QTR.

Faced with several world problems, such as ecological fragility, alpine hypoxia, permafrost, and wind, the construction of QTR has created many "first" at home and abroad:

(1) Tanggula station is the highest railway station in the world. Built in August 2004, Tanggula station is located in the permafrost area of Tanggula mountain at an altitude of 5068 m, covering an area of about 77,000 m^2.
(2) Fenghuoshan tunnel is located on the Fenghuoshan at an altitude of 5010 m, with a total length of 1338 m and the height above sea level of 4905 m on the rail surface. It is all located in the permafrost of the permanent plateau. It is currently the world's highest altitude permafrost tunnel across the permafrost region.

D. Chen and R. Cheng, *Intelligent Processing Algorithms and Applications for GPS Positioning Data of Qinghai-Tibet Railway*, https://doi.org/10.1007/978-3-662-58970-0_2

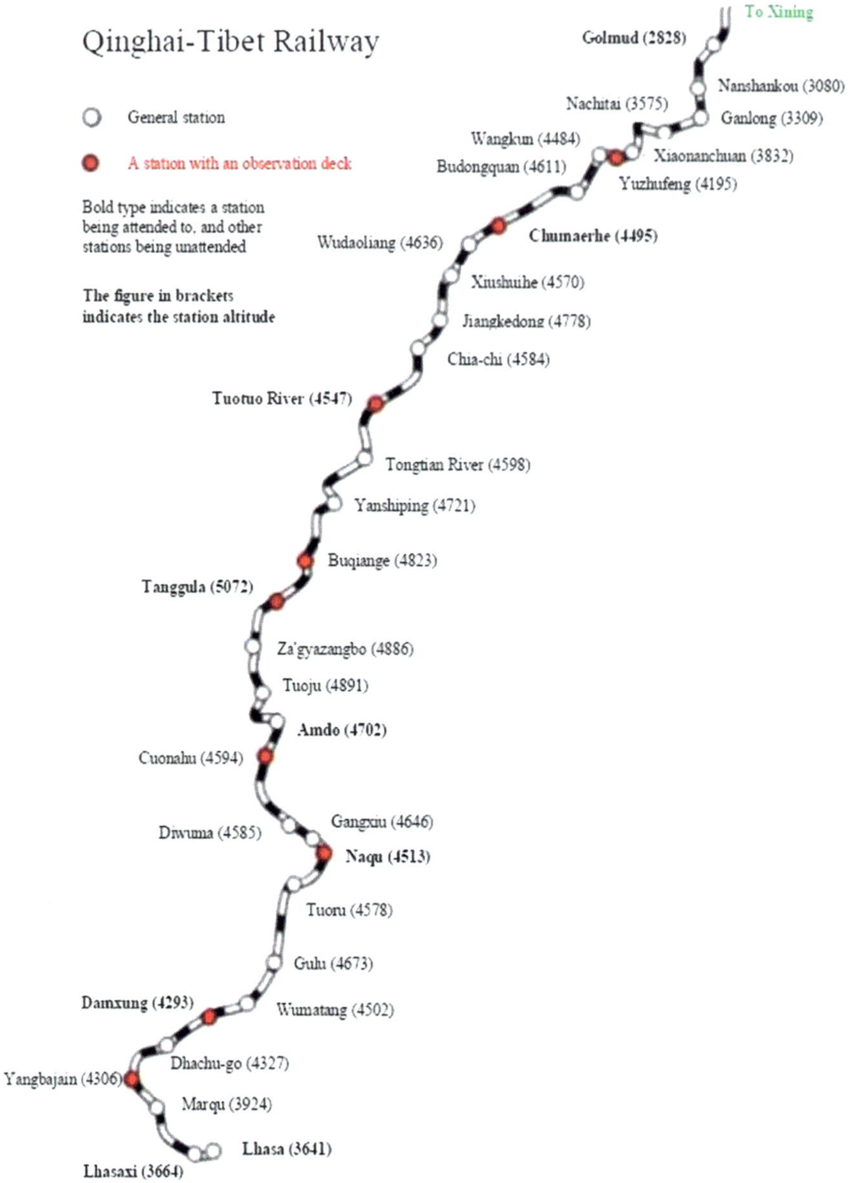

Fig. 2.1 Schematic diagram of the line between Golmud and Lhasa of QTR

(3) Kunlun mountain tunnel, the world's longest plateau permafrost tunnel. The Kunlun mountain tunnel is 1686 m long with an altitude of 4648 m. The tunnel was completed on September 25, 2002.

The QTR is the highest altitude and longest distance railway in the world. The railway on Tanggula mountain has the highest altitude of 5072 m and is the highest railway in the world. Tanggula station is 5068 m above sea level and the highest railway station in the world. Among them, 960 km railway is located at a location of over 4000 m and over 550 km railway is constructed on permafrost area for many years. Built on the Qinghai–Tibet plateau, QTR is one of the most arduous projects among the four major projects of China's Western development strategy. Permafrost, alpine hypoxia, and environmental protection are three major technical problems. With the joint efforts of the Chinese Academy of Sciences, China Academy of Railway Sciences, the railway construction department, and the scientific research and engineering technicians of relevant universities, the three technical problems were successfully solved.

The Qinghai-Tibet railway has the following characteristics: complex geological structure, frequent geological movement, high altitude, strong sunshine, and complex permafrost characteristics. With a total length of 1338m and a rail surface elevation of 4905 m, the Fenghuoshan tunnel is all within the permafrost layer and the ice thickness is more than 150 m. It took only five years to build such a railway in China.

2.2 Permafrost

The frozen soil expands in volume in a frozen state and shrinks in summer. Under the repeated action of these two phenomena, the foundations of roads or houses would break or collapse. The complexity and uniqueness of the permafrost are unparalleled in the Tibetan plateau due to its low latitude, high altitude, intense sunshine, and frequent geological tectonic movements.

When railway is built on permafrost, the roadbed is unstable with the change of temperature. Due to the temperature in Qinghai–Tibet plateau changes greatly, the highest temperature is 38 °C in summer and the lowest temperature is −40 °C in winter [4]. In the season with high temperature, the frozen soil melts, forming hot melt lakes and rivers, the roadbed turns over and slides, forming washboard road. When the temperature decreases, the roadbed freezes and even expands unnaturally, forming the frozen expansion ball [5, 6]. Permafrost contains soil ice, full frozen soil, crack ice, sandstone, mudstone, and sediment interbed. When the temperature rises, the thermal melt expands, especially when the open tunnel is excavated, the slope fails, the landslide, and the base is muddy. After the tunnel is excavated, the arch is seriously broken down and even collapsed. After the tunnel is operated, repeated freeze-thaw will damage the structure and affect the operation safety [7–9]. After railway is opened to traffic, there must be a lot of waste heat discharged from the train, which will affect the railway embankment. To solve the problem of frozen soil, railway scientists in China have been doing scientific and technological research for more than 40 years. In the early 1960s, the Ice Institute of the Chinese Academy of Sciences and the plateau institute of the former ministry of railways investigated the natural environment and permafrost characteristics between Kunlun mountain and Tanggula mountain. For decades, scientists have carried out experimental studies on plateau meteorology, permafrost temperature field, permafrost thermal science, permafrost mechanics, and permafrost construction, bridge construction and tunnel construction.

The temperature of the permafrost region in the Qinghai–Tibet railway plateau has changed greatly in the past 30 years due to the influence of environmental factors such as climate change. Compared with the permafrost in other areas of the world, permafrost thickness of plateau permafrost regions in the Qinghai–Tibet railway is larger, annual average temperature is about −2.1 to −3.4 °C, and the annual average absolute negative temperature is about −5.3 to −7.6 °C. Generally speaking, the temperature fluctuation of 10–20 years can only affect the medium and shallow ground temperature, while the temperature fluctuation of more than 30 years can affect the deep ground temperature. The changes of temperature fluctuation range, time, and space in the Tibetan plateau over 20 years are not enough to cause significant changes in the distribution and thickness of a large area of permafrost. They can only have a significant impact on the surface temperature and permafrost over a depth of 20 m.

Rigorous scientific research enables us to get a general understanding of the changes of frozen soil and find out the various technologies to stabilize the frozen soil. The experts adopted the following methods to ensure the stability and durability of the roadbed:

(1) Adopt schist ventilation roadbed, schist ventilation shield, laying thermal insulation materials, and using hot rod technology. The ventilation roadbed and the ventilation shield make the air convection fast, make the roadbed temperature consistent with the surrounding air temperature, and thus it is difficult to form local heat area, which is conducive to the stability of roadbed.
(2) As shown in Fig. 2.2, in permafrost and unstable places, laying the railways on bridges is adopted instead of constructing railways on the ground, such as Qingshuihe grand bridge.
(3) Waterproof insulation layer is set in the lining of tunnel engineering.
(4) Restudy and formulate technical standards for concrete durability to improve the durability of concrete structures. In order to prevent turtle grain on bridge piers caused by thermal expansion and cold contraction, concrete and frozen soil are "close contact," and negative temperature maintenance measures are adopted. Digging a well and taking wind cooling measures to keep the temperature around 10 °C in the summer. Measures such as baking heat, adding antifreeze, and

Fig. 2.2 Large-scale adoption of "bridging road" in permafrost regions

wrapping the bridge piers with quilts will ensure the durability and anti-freezing of the concrete in the winter.

Now, the bridge culvert and roadbed of QTR have been completed. After several tests of freezing-thawing period, according to the observation data, the original permafrost environment has been protected. Experts attending the 6th international conference on permafrost engineering believe that the engineering measures adopted for the construction of QTR are feasible and its technology has reached the world's advanced level.

2.3 The Harsh Environment of Tibetan Plateau

Along Qinghai–Tibet railway, the area at an altitude of 4000 m above is 960 km, accounted for 84% of the total length of QTR, and the temperature is below −10 °C in many places all the year round. People often say, "when you come to Kunlun mountain, you are already dying. After passing through Wudaoliang, crying father and mother; Climbing Fenghuo mountain, the soul is gone." For every 100 m rise in altitude, temperatures drop by 0.6 °C, and for every 1000 m rise in altitude, oxygen decreases by 10%. At high altitudes, snow is common in summer. In winter, the weather is so cold that the ground is frozen and it's freezing cold all day. At an altitude of more than 4000 m, people often feel dizzy, nauseated, and weak as if they were walking on a ball of cotton. For the lack of oxygen in the thin cold air, the person can feel headache, chest distress, and insomnia, which can cause the disease such as encephaledema, pulmonary edema [10].

In recent years, in order to demonstrate the adaptability of QTR in the future, the National Climate Center of the China Meteorological Administration and the Institute of Environment and Engineering of the Cold and Arid Regions of the Chinese Academy of Sciences studied the century-old climate change in Qinghai–Tibet region through the simulation results provided by the IPCC data distribution center. The results show that, when the impacts of greenhouse gases are only considered, the average annual precipitation in Qinghai and Tibet increases in the mid-twenty-first century, with an increase range of 2.5–10 mm/mon; considering the impact of sulfide aerosols, in the mid- and late twenty-first century, except for the slight increase in precipitation in the north of Qinghai–Tibet region, the precipitation in most other regions will basically decrease, especially the precipitation in southern Tibet will decrease by 5–10 mm/mon, and the precipitation along the railway will increase by 7.5 mm/mon. The increase in the lowest temperature in winter and highest temperature in summer was greater than average. By analyzing the spatial distribution of the temperature change throughout the Qinghai–Tibet area over the next 50 years, if only consider the factor of greenhouse gas increase, the average of year, the winter, and the summer are 2–2.5 °C/50 a, 2–3 °C/50 a, and 2–2.5 °C/50 a, respectively; when considering the common effect of greenhouse gases and sulfide aerosol, the annual average, summer average, and winter average are 1–2 °C/50a, 1.5–2/50 a, and 1.5–2 °C/50 a [11, 12].

No matter how well designed the project needs to be constructed, the Qinghai–Tibet plateau presents many difficulties for construction personnel. The high cold and lack of oxygen seriously threaten the construction personnel of QTR. Usually, people only care about the cold and lack of oxygen in the train after it is opened, but less about the millions of constructers who work in the open air. However, the problem of cold temperature and lack of oxygen does not pose a great threat to the passengers in the train after it is opened to traffic. Just like on an airplane, this problem for the inner train can be solved by taking similar technology and equipment on the train. But for the constructers of the QTR line over construction years, there is always a deadly threat. Experts say that at an altitude of more than 4000 m, it's like carrying more than 20 kg in motion. Therefore, laying a life safety net on the snow plateau, the decision-makers of QTR have developed a rigorous and scientific health security system from the beginning, to solve the problem of hypoxia in alpine regions and ensure the health of the construction workers. The specific measures adopted are as follows:

(1) A set of strict measures have been established for the preventing treatment of plateau disease. It has established a long-term working mechanism of "three defenses" (against plateau disease, plague and epidemic diseases) and improved the medical security system.
(2) Establish the physical examination system. Before working on the plateau, the worker must take physical examination, and the people who do not meet the

conditions are not allowed to work. After working on the plateau for a period of time, the worker also must take physical examination, the worker that is not adapted to the work on the plateau must go back to work on the plain.

(3) Provide enough drinking water for constructors. In view of the high salt content, turbidity, and inaccessibility of drinking water on the plateau, China Railway Construction Corporation and other military medical experts specially developed water purification equipment. The device supplies drinking water for about 7000 people a day.
(4) Establish medical measures such as oxygen inhalation and eating anti-hypoxic drugs. Oxygen is indispensable to human beings. On the plateau, especially on the plateau above 4000 m, people take oxygen for more than one hour every day and often take some anti-anoxia drugs, which have a good effect on protecting health and preventing the occurrence of plateau diseases. For this purpose, China Railway 20th Bureau et al. have developed large-scale oxygen stations, which can not only supply sufficient oxygen, but also diffuse oxygen directly to the construction tunnel. The oxygen diffusion to the tunnel is equivalent to reducing the altitude of working surface by 1200 m.
(5) Acclimatization and stepwise adaptation. After being acclimated in the place below 3000 m for a week or so, the workers will be better acclimated over 4000 m above sea level, and the incidence of altitude sickness will be significantly reduced in the construction personnel.
(6) Adhere to the physical examination system before, during, and after work and construct the system of rest and rotation. Working on the plateau for about two months, the worker will go to a place below 3000 m for a few days to rest, which has a good effect on the recovery of physical fitness. By implementing this rule, the annual rotation of some non-key workers can reduce long-term harm to human body and reduce the incidence of plateau diseases.
(7) Strengthen the system of inspection and night inspection. Early detection, early diagnosis, and early treatment of plateau disease can play a timely treatment of acute altitude sickness, thereby reducing the incidence of acute altitude disease.
(8) Limit labor time and intensity. As the elevation increases, the labor intensity and time of worker need to decrease accordingly. If a tunnel is built at an altitude of more than 4500 m, the working time shall not exceed 4 h per day, and the working time shall not exceed 2 h each time, besides diffusing oxygen directly to the construction tunnel by oxygen supply pipe and carrying oxygen cylinder in the construction site.
(9) Provide good living and living conditions. The bedroom of worker has oxygen, central heating, living area per capita amounts to 2.5 m^2 above, which can prevent the happening of cold and epidemic disease effectively. Ensuring daily supply of fresh vegetables, fruits, eggs, meat can strengthen physique and prevent disease.

In addition, considerable investment has been made in food safety and the prevention and control of plague to prevent the outbreak of plague and epidemic diseases.

Perfect and scientific safeguard system makes the construction of QTR create the amazing miracle of the world plateau operation, that is the zero-death rate of plateau disease. At the international conference on plateau medicine held a few years ago, the experts who attended the conference said that the health security and rescue prevention methods were effective, the personnel's health security was scientific and reliable, and China made contributions to the world's plateau medicine. Over the past four years, tens of thousands of people have gone up and down the Tibetan plateau, fighting in what is known as the "third pole of the world" of life exclusion zone. So far, there has been no case of death due to plateau brain edema, plateau pulmonary edema, and other plateau diseases.

2.4 Plateau Ecological Environment Protection

Twenty million years ago, the young Tibetan plateau rose from the vastness of the ancient Tethys Sea to become earth's closest continent to the sun. As QTR passes through the Tibetan plateau, whether the QTR will be able to protect the last clean land on earth from pollution is another key indicator of the project's success.

Qinghai–Tibet plateau is the birthplace of the Yangtze river, Yellow River, Brahmaputra river, and other large rivers in China and even south and southeast Asia. It is awe-inspiring to see the towering mountains, the vast meadows, and the bleak desert unseen anywhere else on earth. However, the environment of the Tibetan plateau is fragile. If a little careless, it will be destroyed. A careful scientist observed that, the place where they had accidentally dug up a piece of low plant growth 40 years ago has not yet come back and there are no plants growing again [13].

The ecological environment of the Qinghai–Tibet plateau is affected by high, cold and drought in three directions and natural geographical environment. Qinghai–Tibet railway contains four main landscape ecological types from north to south, such as Grey brown desert (south margin) arid shrub landscape area of the Qaidam Basin of Golmud-Kunlun mountain section, northern Tibet alpine grassland landscape areas of Kunlun mountain-Tunggula mount section, Northeast Tibet alpine meadow landscape area of Amdo-Naqu section and South Tibetan mountain and valley shrub grassland landscape area of Damxung-Lhasa section. The wonder of the Tibetan plateau landscape lies not only in its unique geographical environment, but also in the integrity of its original ecology.

We should not only erect the golden bridge of happiness for Tibetan people, but also take good care of the clean and beautiful holy land. From the first day of the construction of the QTR, it has become a new challenge to the construction of the Qinghai–Tibet plateau. The impact of QTR project on the ecological environment presents the following characteristics [14–18]:

(1) Large impact ecological environment areas. The horizontal influence range of the general railway engineering construction on the surrounding ecological environment is limited, and the influence range is usually 300–500 m, which can be

neglected in comparison with the vertical length of the railway. However, for the QTR, its ecological environment is fragile and has a wide range of impacts, which can even affect the ecological environment outside the plateau.

(2) The ecological environment involved is fragile. The QTR passes through the alpine grassland, meadow, desert, and other special and fragile ecological type areas as well as the habitat and breeding ground of wild animals. It is difficult to restore and rebuild the fragile ecological environment of the plateau after being damaged.

(3) Permafrost is very sensitive to railway engineering construction disturbance and often causes adverse geological phenomena.

The unique natural conditions and abundant natural resources of the Qinghai–Tibet plateau have created a rich and colorful flora and fauna, which has been listed as the most important area for global biodiversity conservation by World Wildlife Fund, and also as the priority area for biodiversity conservation action plan of China. The QTR passes through the area where the animal species is few, but the rare species is many and the population quantities of rare species are big. There are about 16 species of mammals, 11 of which are endemic to the Qinghai–Tibet plateau. There are about 30 species of birds, 7 of which are endemic to the Qinghai–Tibetan plateau. The animals that belong to national first-level protection mainly include Tibetan antelope, Tibetan wild donkey, wild yak, white deer, snow leopard, Tibetan snow chicken, black-necked crane, etc. There are about 202 species of 102 genera and 84 endemic species in Qinghai–Tibet plateau.

Tibetan antelope is a unique alpine meadow animal in China. Figure 2.3 shows the Tibetan antelope crossing the bridge tunnel. It has a special habit of regular migration. How to reduce the disturbance to the Qinghai–Tibet railway which lies on the plateau from north to south? The railway command has repeatedly consulted with zoologists to develop conservation plans. Li Jincheng, the chief designer of the Qinghai–Tibet railway, told us that 33 wildlife corridors were designed along the route, some built high bridges to allow them to pass under them, and some built gentle slopes to facilitate their migration. Will the Tibetan antelope buy our bill? There is no answer in the beginning. To find out the answer, scientists have installed night vision monitoring devices in several places. Soon, the news of electronic surveillance came that hundreds of Tibetan antelopes were running east through the bridge, and it seemed that they had gradually adapted and accepted the new changes. A reporter also saw several wild animals that looked like antelopes near the railway line of Yuzhufeng railway station.

Vegetation on the plateau is also very precious. The construction unit requires that the vegetation with humus soil should be moved to other places for preservation, and the vegetation should be moved back the completed embankment slope or the completed site surface after the roadbed is repaired. In the Qinghai–Tibet railway line, the builders also carried out artificial grass experiments. For example, the construction unit has planted 90,000 m^2 of grass in Cuona lake. After the Tuotuohe station, the roadbed slope is carefully planted and transplanted fine green grass [19, 20]. Some people say that QTR is a green road and an ecological road.

Fig. 2.3 Tibetan antelope on the Tibetan plateau

During the construction of Qinghai-Tibet railway, environment protection facilities are designed and put into operation at the same time. The environment protection facilities are used to effectively protect the permafrost environment, the water quality of river, wildlife migration and the natural landscape on both sides of railway.

The basic content of environmental protection of QTR has two aspects: pollution prevention and control ecological protection. Pollution prevention and control mainly include the prevention and control of water pollution, garbage pollution, and air pollution [21, 22]. The QTR passes through the sources of the Yangtze river, Yellow River, and Lancang river. Environmental protection focuses on ecological protection, especially the protection of wildlife, vegetation, wetland system, water source, nature reserve, and natural landscape:

(1) Protect permafrost environment and plateau vegetation do the grass planting experiment. Strictly optimize the site selection of temporary land, control the floor area, minimize the construction of houses and roads in areas with vegetation, and limit the activity scope of construction personnel and mechanical vehicles. Vegetation transplantation and topsoil preservation are carried out. In other words, turf or topsoil should be shoveled and stored in other places in the construction. After completion of the project, it should be paved back to the original place to promote vegetation restoration or lay the foundation for veg-

etation restoration. In order to solve the technical problems of ecological and environmental protection and recovery in the alpine environment of the plateau, experimental studies were carried out on vegetation restoration of the plateau, vegetation construction technology of roadbed slope, vegetation restoration, and reconstruction of alpine meadow, all of which achieved periodic results. The survival rate of sod transplantation is very high in Fenghuoshan and Tang-Nan section. The vegetation restoration and reseeding experiment of roadbed slope in different climate zones, such as Dangxiong, Ando, and Tuotuo river, were successful.

(2) Protect wildlife. On the traditional migration routes of wild animals, 33 special wildlife corridors are set up, and the cumulative length of the road along the railway is nearly 60 km. During the migration season of Tibetan antelopes, the construction of Tibetan antelope should be stopped to make way for the migration of animals. To properly handle production and living waste, actively prevent soil erosion, and strictly control the occupied area of the surface.

(3) Protect wetland and water resource environment. In order to ensure that the ecological function of the river source area is not affected, the wetland ecological environment and the route of the wetland area are effectively protected. In order to ensure the supplement of water resources to the wetland by surface runoff and the water conservation function, prevent the wetland from shrinking, the concrete measures are as follows: constructing bridge culvert in gully, increasing the density of small bridge culvert, using bridge instead of constructing the roadbed on the ground and filling in seepage materials in subgrade, etc. For the section along the river and lake, a special eco-environmental protection construction organization and design scheme has been formulated to prohibit hazardous materials, such as garbage and construction waste, from piling up near the river and channel water bodies. After concentrated precipitation treatment, production wastewater and domestic sewage are used for greening or dust removal. It is strictly prohibited to be discharged into river water body, which effectively prevents pollution to water resources and wetlands.

(4) Reasonable treatment of production and household waste. Biodegradable and non-biodegradable waste is classified and treated along the route. Biodegradable waste is filled in the site, and non-biodegradable waste is transported to the dump under the mountain for treatment.

(5) Prevention of soil erosion. Strictly control the disturbance area to the surface, the disposal of soil and waste residue should follow the principle of first treatment and then discard.

To sum up, through the construction of the "four-in-one" environmental protection management system, professional monitoring institutions were entrusted to monitor the river water quality and soil erosion during construction. Meanwhile, local environmental protection administrative departments signed environmental protection responsibility letters with construction units to comprehensively supervise, monitor, and supervise the implementation of various environmental protection measures. Insist on scientific and technological innovation, actively organize and carry out scientific research on environmental protection, absorb the opinions and suggestions of experts from various aspects, and largely adopt new technologies and new techniques in the construction to overcome the environmental problems encountered in the construction of Qinghai–Tibet railway [23].

2.5 The Development of the Qinghai–Tibet Railway

The QTR is a landmark project of China's Western development strategy. Since opening, China Railway Corporation and Qinghai and Tibet provinces are committed to "mining the development potential of QTR to the greatest extent, and playing a powerful radiation role of QTR," which can vigorously promote two provincial scientific and harmonious development of economy and society, and the QTR has become the economic line of the Tibetan people.

The QTR greatly shortens the distance between the snow plateau and the inland, greatly facilitates the travel of people of all ethnic groups in Qinghai–Tibet as well as the entry of tourists from various regions, and expands broad space for economic and social development. In 2007, the GDP of Qinghai–Tibet region reached 34.219 billion yuan and increased by 14%. Qinghai's GDP reached 760.96 billion yuan and increased by 12.5%, which is the largest increase in 10 years. These are closely related to the operation of the QTR.

The low operating cost of QTR has created good economic benefits for enterprises along the railway and benefited people of all ethnic groups. The freight rate of railway transportation is 0.12 yuan per ton per kilometer, which is 0.27 yuan lower than highway transportation. Based on the freight rate of this unit, the transportation cost of goods between Xining and Lhasa can be reduced by 528.12 yuan. The reduction of transportation cost has improved the competitiveness of Qinghai–Tibet commodities. At the same time, a large number of high-quality and inexpensive commodities entered Tibet from the mainland, not only changing and enriching the consumption structure of fuel and food for the people of all ethnic groups, but also reducing the consumption expenditure of residents and improving the quality of life. As shown in Fig. 2.4, comrade Hu Jintao, then President of China, attended the opening ceremony of the second phase of the QTR.

Fig. 2.4 Opening ceremony of QTR

Qinghai–Tibet railway plays an important role in transforming the Tibetan plateau from resource advantage economy to economic advantage economy. In 2007, more than 4 million tourists visited Tibet. At present, tourism has become an important pillar industry in Tibet and has stimulated the development of hotel services, catering services, retail and wholesale of tourist supplies and other industries. The annual production capacity of brand "5100 glacier mineral water" of Tibet Glacier Mineral Water Co., Ltd. is 300,000 tons. Located in Golmud, Qinghai Salt Lake Group is rich in potash resources and the railway creates favorable conditions for the rapid development and expansion of the company.

Since its opening, the Qinghai–Tibet railway has not only transported a large number of tourists from home and abroad to Tibet, but also facilitated the people of all ethnic groups in Qinghai and Tibet to go out, further expanded the channels of cultural exchanges, expanded the cognitive range of Tibetan culture, and promoted the harmonious development of Western traditional culture and modern civilization.

Railway construction and development on the Qinghai–Tibet plateau are still progressing. According to the national medium and long-term railway network planning,

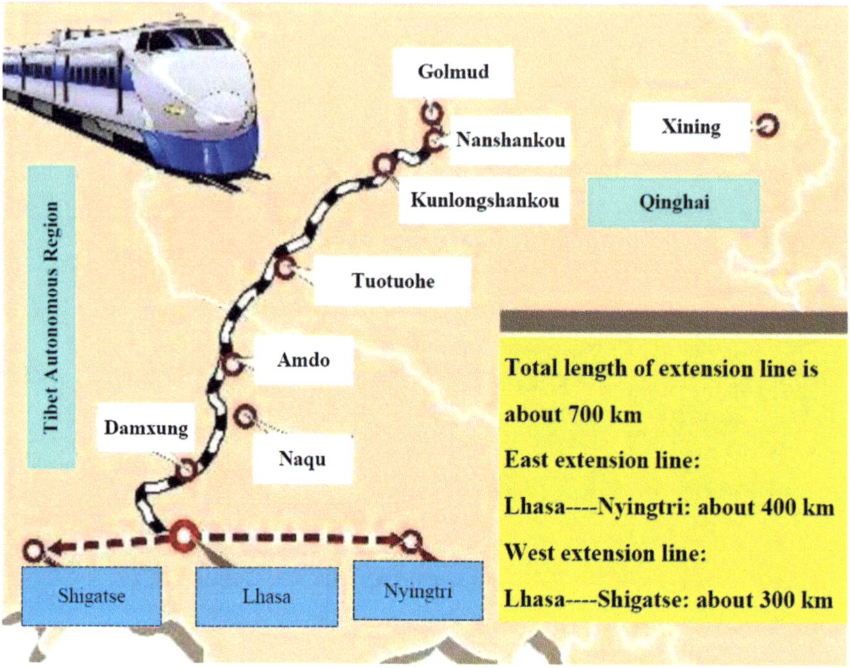

Fig. 2.5 Planning map of QTR

China will use the Qinghai–Tibet railway as the link to build six trunk railways from Xining to Zhangye, Geermu to Dunhuang, Geermu to Korla, Geermu to Chengdu, Geermu to Xigaze, and Lhasa to Nyingchi, as shown in Fig. 2.5. The Qinghai–Tibet plateau will have closer economic and cultural ties with all parts of the country. As an important national infrastructure, railway has a strong "hematopoietic function," which will play a significant role in promoting the Western development and promoting regional coordinated development.

References

1. Wu QH, Tong CJ. Change of the permafrost and stability on Qinghai-Tibet Road. J Glaciol Geocryology. 1996;17(4):350–5.
2. Xi XL, Xu ZY. Ecological environment protection in the construction of Qinghai-Tibet railway. Environ Sci Technol. 2005;28:119–21.
3. Liang CG. Qinghai-Tibet railway: a century project that benefits all ethnic groups in Qinghai-Tibet. Chin Railways. 2008;11:7–9.
4. Li DL, Guo H, Wang W, Wei L. Prediction of annual average temperature change along Qinghai-Xizang railway. Plateau Meteorol. 2003;22(5):431–8.
5. Zhang LX. The ground temperature variation law of the plateau frozen soil area of Qinghai-Tibet railway and its influence on the stability of roadbed. China Railway Sci. 2000;21(1):36–47.

6. Bai HZ, Li DL, Dong AX, Fang F. Strong wind and wind pressure along the Qinghai-Tibet railway. J Glaciol Geocryology. 2005;27(1):111–6.
7. Zhou ZX, Wei QZ, Xu M. Experimental analysis of station roadbed and common width of subgrade in permafrost region of Qinghai-Tibet plateau. Railway Stan Des. 2006;1:9–12.
8. Xu L. Analysis of geothermal characteristics for gravel slope protection and stone protection road base in permafrost region of Qinghai-Tibet railway. South-West Jiaotong University; 2006.
9. Liu D. Analysis of engineering effects of slice stone air-cooled embankment on lower temperature permafrost regions in Qinghai-Tibet railway's operation period. Railway Constr Technol. 2007;4:58–63.
10. Xu Y, Ding YH, Li DL. Climatic change over Qinghai and Xizang in 21st century. Plateau Meteorol. 2003;22(5):451–7.
11. Tang MC, Zhong HL, Li DL. The standard of marking off the four seasons along Qinghai-Xizang railway and its temperature variation analysis. Plateau Meteorol. 2003;22(5):440–4.
12. Emori S, Nozawa T, Abe-Ouchi A, et al. Coupled ocean-atmosphere model experiments of future climate change with an explicit representation of sulfate aerosol scattering. J Meteorol Soc Jpn Ser II, 1999;77(6):1299–1307.
13. Chen H, Li SC, Zheng D. Features of ecosystems alongside Qinghai-Xizang highway and railway and the impacts of road construction on them. J Mt Sci. 2003;21(5):559–67.
14. Zhang YQ. Negative impact of Qinghai-Tibetan construction on ecological environment of Qinghai-Tibetan Plateau. Bull Soil Water Conserv. 2002;22(4):50–3.
15. Wang MZ, Xu ZY, Yang CY, Sun ZK. Effect of Qinghai-Tibet railway construction on plateau eco-environment. Environ Prot Transp. 2002;23(3):2–4.
16. Sun SY. Ecological environment characteristics and protection measures along the Qinghai-Tibet Railway. Railway Occup Saf Health Environ Prot. 2002;29(3):121–4.
17. Zhang Y, Shang JC. Discernment of strategic environmental impacts on the decision making of Qinghai-Zang railway. Environ Prot Xinjiang. 2001;23(3):24–8.
18. Chen XD. Analysis of the impact of Golmud-Lhasa section of Qinghai-Tibet railway on environment. J Beijing Univ. 2002;24(5):135–8.
19. Liu L. Study on landscape ecology method of ecological environmental impact assessment of traffic engineering. Southwest Jiaotong University; 2006.
20. Chen GC, Zhou GY, Sun J, Chen ZG, Han YJ, Lu XF. Application of Kengyilia thoroldiana to vegetation restoration in gravel-soil-taken field along the Qinghai-Tibet railway. J Glaciol Geocryology. 2006;28(4):506–11.
21. Ran L. Design and research of the railway on Qinghai-Tibet Plateau. China Railway Sci. 2001;22(1):16–22.
22. Gu XA. Preliminary study on the effect of railway noise on bird habitat and reproduction. Railway Lab Saf Health Environ Prot. 1999;26(2):82–4.
23. Hou SQ, Li DS, Li J. Experimental study on the treatment of low temperature Sewage on Qinghai-Tibet railway line. China Railway Sci. 2005;26(3):104–8.

Chapter 3
Incremental Train Control System of Qinghai–Tibet Railway

3.1 Train Positioning Methods

The application of GPS technology in Qinghai–Tibet railway mainly focuses on the navigation, monitoring, command, and dispatching system of the train positioning. GPS is used as the auxiliary train positioning device in the railway train control system of Qinghai–Tibet railway and is taken as a backup of the query balise (redundant system) in the moving block control system. The vehicle-mounted GPS shown in Fig. 3.1 is used as the auxiliary correction of the reference position of the ground track. GPS-aided dispatching system uses GPS to compute the actual positioning, the operational direction and speed of the operational train, and then transmits the positioning results to the dispatch center through wireless communication channel to realize the real-time whole-process monitoring and ensure the safety of the train. The continuous operational location information of train provided by GPS covers the whole operational line of the train (station and zone), which can be used to adjust the operation plan, train dispatching and controlling, and accurate real-time monitoring.

The functions of GPS positioning system:

(1) Map Display

Many vectorgraph and raster graphic be opened at the same time and the same map can be also opened for many times, so that different maps for different regions and different parts of the same map can be shown. The panorama of operational track of train can be displayed through a map created by GPS.

(2) Train Position Display

The location of each train is displayed on the screen with a specific symbol, and the operator displays it in either of three ways. Constantly display (the default display mode), all operating vehicles of the same team display in the close-up area according to the operator instructions, the same train doesn't display according to the operator instructions until the alarm is received.

D. Chen and R. Cheng, *Intelligent Processing Algorithms and Applications for GPS Positioning Data of Qinghai-Tibet Railway*, https://doi.org/10.1007/978-3-662-58970-0_3

Fig. 3.1 The vehicle-mounted GPS device

(3) Map Operations

Basic map operations include amplification, narrow, roaming, hierarchical display. In addition, show the latitude and longitude of an arbitrary point on the map also, calculate the distance between any two points, query the geographical target information superimposed on the map.

(4) Simulation Features Based on Real-Time Positioning Data

The known target position information can be entered into the computer and displayed.

(5) Real-Time Positioning Data Display

The real-time position of the moving target can be displayed as a frame list, such as longitude, latitude, speed, state, time, and date.

(6) Historical Positioning Data Display

Operator can display the history operational trajectory of the related trains according to user's need, deposit or bring up operational data of each train, such as identification number of vehicle system unit, vehicle type and vehicle model, the driver information, etc.

3.2 Communication Modes

Because of the bad operational environment of Qinghai–Tibet railway (QTR), advanced communication technology should be applied to QTR. GSM for Railway (GSM-R) communication system has a great advantage for railway besides all functions of GSM. GSM-R communication system applied to QTR can meet the plateau railway demand for communications, realize the seamless communication for 160–500 km/h high-speed trains, obtain improved command, and dispatch function. The structure diagram of GSM-R communication network of Qinghai–Tibet railway is shown in Fig. 3.2 [1].

3.2.1 GSM-R Features

Based on GSM cellular system, GSM-R is a wireless cluster communication system, which increases the function of dispatching communication and is suitable for practical elements in high-speed environment. The working frequency range is 900 MHz, the transmitting frequency of mobile terminal is 885–889 MHz (base station receiv-

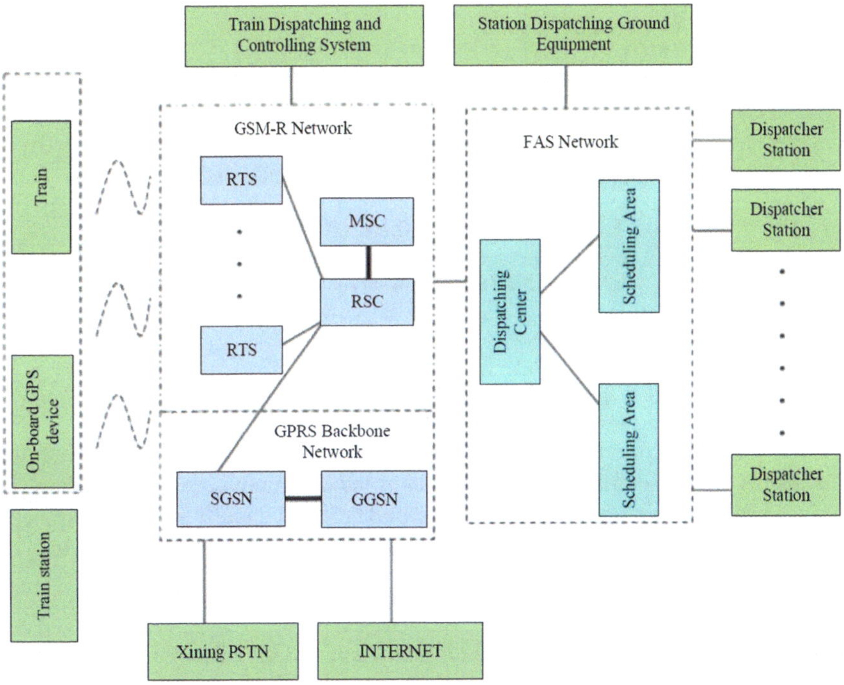

Fig. 3.2 Structure diagram of GSM-R communication network of Qinghai–Tibet railway

ing), the transmitting frequency of the base station is 930–934 MHz (mobile terminal receiving), the channel bandwidth is 200 kHz, and the duplex frequency separation is 45 MHz.

3.2.2 GSM-R Functions

(1) Dispatching Communication Function

The communication service of dispatching communication system includes train dispatching communication, freight dispatching communication, traction substation dispatching communication, other dispatching and dedicated communication, station-yard communication, emergency communication, construction maintenance communication and crossing communication, etc.

(2) The Transmission of Train Number and Train Stabilized Information

Train number and train stabilized information have the vital practical significance for railway transportation management and train safety. The information transmission can be achieved by data acquisition and transmission application system based on GSM-R circuit switching technology or realized by GPRS.

(3) Scheduling Command Transmission

The railway dispatching command is a written order issued by the dispatcher to the driver. It is an important guarantee for the safety of the train. Using the channel GSM-R system to transmit dispatching command will undoubtedly speed up the delivery of the dispatching command and improve the working efficiency.

(4) The Information Transmission of the Train's Tail Equipment

The wind pressure data of train's tail can be sent back to transmission channel of GSM-R communication system, which can easily solve the problem of wind pressure data transmission.

(5) The Transmission of Shunting Cab Signals and System Monitoring Information.

GSM-R communication system can provide transmission channel for shunting cab signals and the monitoring information of the locomotive, and thus realizes the data transmission between the ground equipment and the multi-vehicle equipment. It also can store the entering and exiting information about the shunting mode.

(6) Train Control Data Transmission

Providing two-way secure data transmission channel, GSM-R communication system is adopted to realize two-way wireless data transmission between the train and ground equipment.

(7) Internal Mobile Service Communication

GSM-R communication system provides the internal communication service among water and electricity, public works, signal, communications, power supply, bridge guard departments. Operation personnel can communicate with station attendant, departments' dispatcher, or autovon user by handheld machine. The operator can also call the driver and establish a call with the driver in emergency circumstances.

(8) The emergency command communication.

The emergency communication, including voice, image, data information, will be established within emergency sites or between emergency sites and rescue center when the natural disasters or emergencies affecting railway transportation occur.

3.3 Incremental Train Control System

The special geographical environment and poor natural conditions of Qinghai–Tibet railway put forward higher requirements for the modernization of railway signals. In Golmud–Lhasa section of Qinghai–Tibet railway, global positioning system (GPS) technology is finally determined to achieve train positioning information. The integration functions of the signal system, including interlock, blocking, and train overspeed protection, are realized by taking the GSM-R communication system as the information transmission platform between the train and ground equipment and choosing incremental train control system created by General Electric Company (GE) as the train control system. ITCS system combines the wheel sensor with the differential GPS positioning system to realize the precise positioning of the train. The train is equipped with ATO-onboard equipment to ensure the safety of the train. GSM-R digital mobile communication platform is used for information transmission between the train and ground equipment. The train station is equipped with the radio block center (RBC) and the Vital Harmon Logic Controller (VHLC), which realizes the interlocking of the station and forms a new train operation control system. Compared with the traditional signal system, this system can not only improve railway passing capability, shorten the train running interval, but also greatly reduce the wayside equipment [2].

On the basis of existing ITCS signal system applied in Michigan line in the USA, ITCS signal system equipped in the Qinghai–Tibet railway is realized by considering both the actual operational conditions of Qinghai–Tibet railway and China Railway Technical Standard based on GSM-R and GPS wireless train control system. Its characteristics include: (1) high reliability and no personal management for small station; (2) simple structure and suitable for harsh external environment; (3) high fault-safety and reliability; (4) Multi-channel robust information transmission channel, including wired and wireless mode, is adopted for building resilience to natural disasters; (5) non wayside equipment used in the middle or small stations and blocking zone, unmanned management for making the operation cost low.

3.3.1 ITCS Functions

Since 1994, the development of ITCS system has been carried out in combination with the accelerated reconstruction project in the northeast corridor of Michigan of the USA. The design target speed is 250 km/h, with a total length of over 200 km. The system started trials of operation in 1998 and was put into commercial operation in 2000. The maximum speed of the line is 126 km/h under the original traditional blocking and has increased to 145 km/h after using ITCS (restricted by line conditions). At the end of 2005, the maximum speed went up to 195 km/h. The system safety certification program has been completed between 2004 and 2005.

The design of Qinghai–Tibet railway line has three specific requirements: passengers travel consumed in the shortest possible time, the maintenance-free equipment, basically "unmanned" management along the railway [3]. To meet these specific requirements, the Qinghai–Tibet railway adopts the ITCS signaling system, which uses the integrated design idea and integrates automatic blocking, station interlocking control, and train overspeed protection control together. For this reason, ITCS can obtain the turnout location, signal display, and track occupancy through the interface with the existing interlocking system. It adopts bidirectional continuous wireless communication and uses the onboard signal as the main control signal for the train operation. By monitoring the train operation in real-time, ITCS system provides warning signal or braking signal for the driver or braking system when the situation changes. The train will be braking to ensure the safety of the train in case of emergency.

ITCS mainly has the following three functions:

(1) Virtual Automatic Block and Automatic Block Between Stations

ITCS divides the interval between stations into several virtual block partitions. The virtual blocking section information will be stored in the radio block center (RBC) of railway station in the form of database (electronic map), so it is called virtual automatic block. The number of virtual block partition can be increased by modifying the software. The system sets virtual semaphore signal to protect each virtual block partition, and the railway semaphore signal can display two kinds of signals: ban signal or allowing signal. Virtual block partition and virtual semaphore signal are set for both upper direction and downward direction. The virtual semaphore signal can be set to juxtaposition and stagger setting, but the ground track circuit and the ground semaphore signal are not used.

Automatic block between stations is a kind of virtual automatic block, which likes ITCS controls the blocking sections by combining several virtual blocking sections into one blocking section. Mode transition needs operator's management, each station sets two buttons (virtual automatic block or automatic block between stations) in the position of both upper direction and downward direction, and the system automatically verifies the transition timing and conditions.

(2) Unattended Station Interlocking Control

To meet "fault-safety" principle, VHLC is used to realize unmanned station interlocking task. VHLC arranges the route for coming train and shunting task by receiving and executing the scheduling command of Centralized Train Control (CTC).

(3) Train Operation Control and Overspeed Protection.

As a safe signal control system, ITCS can ensure the speed limit operation and overspeed protection control under the principle of fail-safe. The communication between ground equipment and onboard equipment is realized by applying GSM-R wireless communication network, by which ITCS can obtain the virtual signal status information from ground RBC and VHLC.

The onboard signal is used as the main signal to control the operation of the train. ITCS follows the principle of human-computer control and human-priority control. In other words, ITCS overspeed protection system does not interfere with the operation of the driver under normal circumstances, only as the ultimate measures to ensure the operational safety of the train.

ITCS has the advantage of observing multiple goals ahead. According to the onboard electronic maps, onboard computer can determine the position of the train by monitoring train's running speed in real time. Based on the speed sensors and GPS receiver, onboard computer can continuously compute the train running speed, the distance to multiple target points ahead and the speed-distance curves for each target speed, from which the most restrictive and strictest mode curve will be selected. The system can provide the driver with the alarm information in the form of sound and light when the train exceeds the permitted speed. If the driver doesn't take measures to reduce speed to allowable value, the onboard computer will trigger punitive braking to ensure driving safety [4].

3.3.2 The Components of ITCS

ITCS system consists of center equipment (including the dispatch center and maintenance center), onboard equipment, and railway station. The data exchange between devices is realized by a digital communication network platform, which includes the digital cable communication network platform and GSM-R digital communication network platform. Figure 3.3 shows the structure of ITCS system.

(1) Central Equipment and Functions

The central equipment mainly includes centralized traffic control (CTC) and maintenance terminal (MMT). Its main function is to give dispatching instructions, receive and process both the running status information and maintenance information.

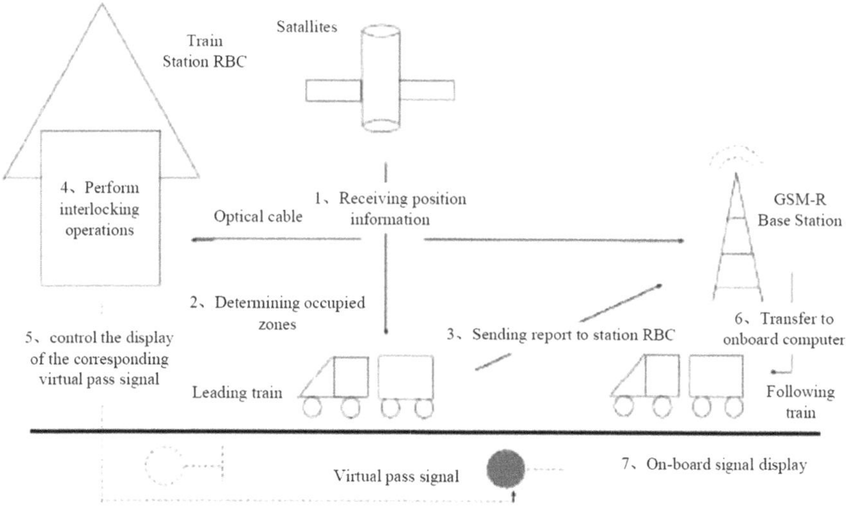

Fig. 3.3 Structure of ITCS system

(2) Wayside Equipment and Functions

The wayside equipment mainly includes VHLC, RBC, GPS differential station, CTC and station self-regulation RTU and local computer-based interlocking (CBI). The wayside equipment receives the scheduling command of CTC, while the feedback signal is expressed. For the local interlocking station, CBI will complete the interlocking function in the station, and VHLC will monitor the interlocking information; for non-local interlocking stations, the interlocking function in the station will be completed by VHLC. The logic operation of the virtual passing signal machine is performed by RBC. The information of CBI, VHLC, and RBC is exchanged to complete interlocking and self-closing functions. GPS differential station provides differential information to the system to improve the positioning accuracy and integrity.

The wayside equipment will also transmit the status of the signal equipment in the jurisdiction area to the onboard system and obtain the train occupancy information through the wireless communication system. The RBC in the original station allows the train to download the database and complete departure test.

(3) Onboard Equipment and Functions

The onboard equipment includes the onboard computer and the rear-end equipment. Based on the information received from RBC, the onboard computer is used to confirm the status and display of the various signal devices in the front of the track and provide overspeed protection for the driver.

The main onboard equipment for wireless communication includes the MSC and ITCS message forwarding equipment, and the wireless communication which provides the message forwarding function for the ITCS vehicle-ground communication

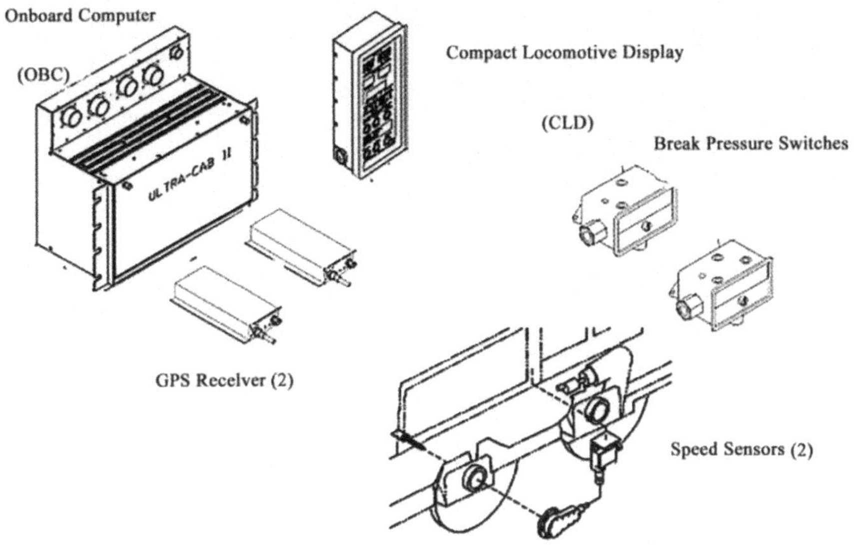

Fig. 3.4 Onboard computer structure

[5, 6]. The bidirectional wireless communication between the onboard equipment and the wayside equipment is realized by GSM-R network. The structure of onboard computer is shown in Fig. 3.4 [7].

3.3.3 The Operating Principles of ITCS

Based on virtual automatic blocking, the control system can send dispatching command directly to the station execution unit and onboard equipment through wired optical communication and GSM-R wireless network. Station execution unit (SEU) executed interlock checks for operational zones and the related virtual sections within the railway station by train-ground communication. SEU checks the resource occupation of operational zones and station sections according to the principle of hostile survey and then registers and allocates the route for the train to ensure the uniqueness principles of interlocking logic.

The information transmissions among onboard equipment, railway station, and dispatch center are realized by GSM-R wireless network. The checking and allocation task of train route is accomplished through the cooperation of both wayside equipment and onboard equipment, which include vital computer and RBC, wayside VHLC, switch machine control module (HSC), dedicated communication router.

Based on the wired and wireless communication mode, the movement authority (MA) is issued to onboard equipment or the railway station by scheduling command when the train runs in the station or the operational zones.

In addition, the occupancy and the readiness state of the idle route, the state of station's virtual signal machine can be transmitted to the train control center through the optical fiber communication. Then, the information received by train control center will be transmitted to the locomotive by GE dedicated routing and the mobile switching center (MSC) of GSM-R network.

The occupancy information, train operational information, train positioning information will be transmitted from onboard equipment to the TCC via GSM-R base station, and then transmitted to station's VHLC for the related safety logic calculation. After the upcoming sections have been safely verified and the dispatch command and transmission address have been checked, VHLC can send control command to the wayside control unit HSC and onboard equipment and allocate the route for the train. At the same time, the onboard can obtain the moving authority of the relevant route. When the routing assignment is successful, moving authority and scheduling commands are sent to the onboard equipment and the driver drives according to the corresponding information displayed by DMI. The ITCS minimizes the number of field equipment, such as the number of wayside signal equipment, the number of track circuits, signal machines, and corresponding control cables at stations and train operation sections. The management of station realizes dehumanization and informatization. Train integrity inspection is realized through the train-tail device (EOT) and GPS, and the train integrity inspection can be carried out without interruption during the whole course of train operation by combining the train-tail device (EOT) and GPS with the wayside device or through the onboard equipment [8].

3.3.4 System Reliability Analysis

The performance index of railway signal system is mainly related to reliability and safety. The mean time between failures (MTBF) is generally used to quantitatively measure the system reliability. In addition, mean time between faults on the dangerous side (MTB-FAS) is used to quantitatively evaluate the system safety.

The ITCS system is composed of multiple modules. In order to facilitate analysis and calculation, the following assumptions are made according to the actual use situation: The fault is independent from each other, and there is no correlation between faults; the mean time to recovery includes the time of fault location, repair, and restart; the faults of system power supply and cables are not considered. The system reliability depends on various subsystems, which are made up of several functional modules. As shown in Table 3.1, the reliability of each functional model can be determined through the reliability prediction.

As shown in Fig. 3.5, the fault tree model of onboard unit for ITCS can be modeled by using fault tree method to analyze the reliability of the system.

According to the characteristic of Markov process, to quantitative calculation and gate, or gate, the system is calculated MTBF of 1.8e−6 h, visible, ensure the reliability of each module according to the requirement, and can make the system have high reliability level [7].

Table 3.1 Reliability of each functional module

Failure modes	Failure rate
HCPU	3e−6
CPU2	3e−6
ILOCK	3e−6
MFIO	3e−6
LP	3e−6
TVO	3e−6
CLD	1e−5
Odometer	1e−5
GPS module	1e−5

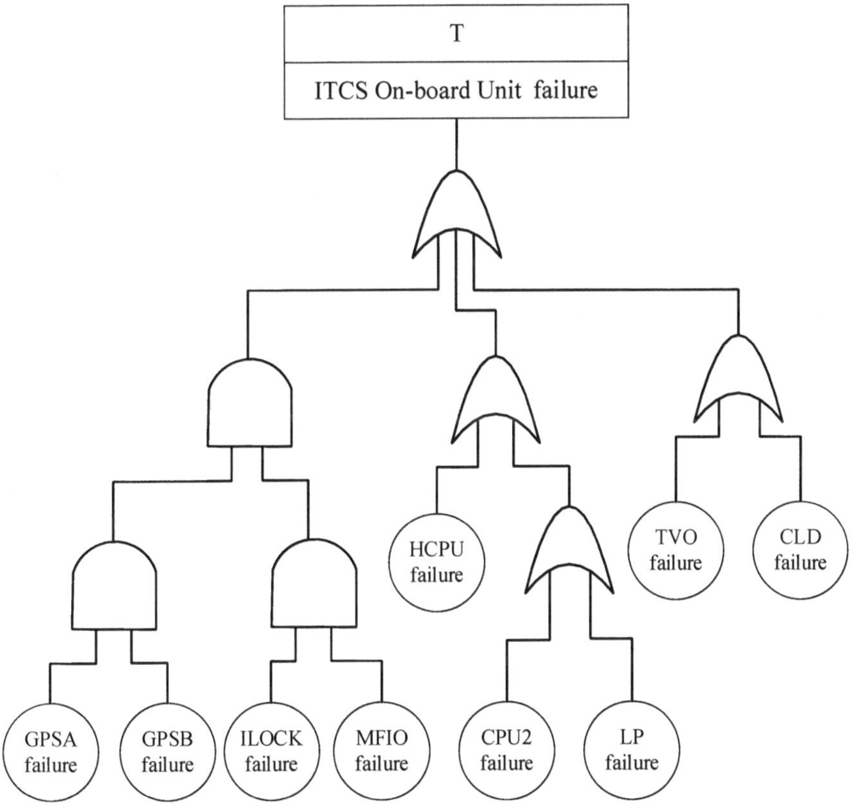

Fig. 3.5 Fault tree of onboard unit for ITCS

3.4 Onboard Database

Onboard database stored in the onboard computer is the basis of ITCS, which contains the position information of all stations, the virtual signal, the wayside signals, radio blocking center, and permanent and temporary speed limit information. The train's operations are determined by comparing the train position with the database data. In addition, the line description information is taken as the most direct basis of map data for train positioning. This stored information describes the immutability of railway line and wayside infrastructures and mainly consists of the following parts:

(1) Railway topographic data is mainly used for train positioning, and the line topographic data should fit the geographical position of the line and match the original line segment in the initialization process of onboard system. The establishment of the database can be provided by the existing electronic map or can be automatically generated by the readout tool according to the line description information.
(2) Railway route map contains position information (latitude and longitude) and curves' parameters measured by GPS. The route map is the basis for the onboard computer software to convert the longitude and latitude position to the ITCS mileage position. High-precision GPS positioning data and altitude graphs are necessary data for train positioning.
(3) The positioning data for infrastructure targets includes both the position of the infrastructure and the length between the targets and must be accurately described.
(4) Line documents describe the line conditions of each section of ITCS, including various signal machines, switch machines, slope, etc.
(5) Speed limit table describes the speed limit data of each section [9].

3.5 Emergency Strategy in Case of Failures

3.5.1 Failure Modes of Onboard Equipment

(1) Onboard safety computer (OBC) fault management. In case of failure of OBC, the system will trigger emergency braking and change the current operational mode to failure mode, inform the driver in time through the onboard driver-machine interface, send the fault report to the station and the dispatch center through the onboard communication equipment, send the failure alarm to the train administrator and send the failure report to the maintenance management terminal.

(2) The reading of speed sensor is inconsistent. When the input data of the two speed sensors is inconsistent and the difference exceeds the predetermined threshold value of the system, the system will start self-inspection program, issue sound and light alarm, and start punitive braking after preset seconds.
(3) The automatic brake truncation pressure switch is not closed. When the onboard computer detects that the automatic truncation pressure switch is in the open state, the onboard computer will switch to the train-tail mode, and the electromagnetic valve of braking will punitively lose power.
(4) The automatic brake truncation pressure switch is not opened. In train-tail mode, if this happens, the train will enter OUT or YARD mode. In addition, the train will be controlled in YARD mode when the train runs within the ITCS area indicated on the onboard electronic map, otherwise it is controlled in OUT mode.
(5) Confirmation button was not pressed in time. When the system's sound and light alarm requires the driver to press the confirmation button, the system will start the acoustooptic prompt and display the 10 s countdown. If the driver fails to confirm within 10 s, the system will start the punitive braking.
(6) The mode switch failed to be switched successfully. When the system prompts the driver to switch the mode switch, the sound and light prompt will be activated and the 10 s countdown will be displayed. If the driver fails to switch the switch to the correct position within 10 s as prompted, the system will handle the switch effectively according to the specific situation.
(7) Onboard radio failed. When this happens, the onboard computer will take the nearest signal machine in front of it as the stop signal. If the signal machine is a home signal machine, it will be shown as an absolute stop. If the signal machine is an interval signal machine, it is displayed as a parking permit. Once the current TTP calculated by the onboard computer has reached or is less than the predefined value, the system will trigger braking.
(8) GPS positioning failed. When the speed sensor is used as the input of positioning calculation, if the positioning error exceeds the threshold set by the system, such as 132.5 m, the system will switch to station mode and implement low-speed protection for the train (e.g., 32 km/h).
(9) GPS time synchronization information lost. The GPS time stamp is lost during operation, and the radio station will stop sending messages after 5 min, which will be processed as a communication failure.
(10) The tail device failed. In the protection mode of the tail device, when the tail device fails, the onboard computer can allow the train to continue to operate normally after stopping the tail detection function. However, the system will no longer clear the interval according to the most unfavorable position of the tail of the train. The system will automatically block all sections the train goes through after the tail device has failed.
(11) Onboard computer software's self-inspection reports an error. The system continuously checks the safety-related variables, and the error of inspection will lead to the magnetization of the brake solenoid valve and trigger the braking.

(12) The onboard computer lost power. When the onboard computer is working in the IN mode, the loss of power will cause the electromagnetic valve to lose power and trigger braking.
(13) Database validation fails or database files fail. If the onboard system is working in IN mode and the failure database file is the database file that describes the line information of the front-controlled area, the last signal machine of the front-controlled area stored in the current effective database file is regarded as a stop signal machine. After the train stops in front of the stop signal machine, it will turn to station mode and run at a speed limit when receiving a overtaking signal. If the failure database file describes the current control area, the onboard system will immediately turn to station mode and run at a limited speed.
(14) The temporary speed limit file is invalid. When the failure is detected, the onboard computer will attempt to communicate with RBC and update the temporary speed limit file. If there is no response after 20 s of failed update, the temporary speed limit file in the control area is invalid, resulting in the processing method equivalent to the database file failure in the control area.

3.5.2 *Failure Modes of Wayside Equipment*

(1) An error occurred in the input of signal representation. When the signal representation of VHLC is wrong, the system will degrade the signal representation.
(2) Input error of the switch machine. When the switch is in a state of loss representation, the onboard computer will trigger the braking to stop the train. If the train passes through the switch, it will enter station mode (because the onboard computer cannot determine the current working state of the track). The on-board computer can begin to re-evaluate the working state of the current track only when the train passes the next signal device.
The system may not be able to monitor the switch when it is blocked or the switch control equipment fails. At this point, in the case of ensuring route safety of the train, the emergency control panel of train traffic control can be used to implement the single operation control of the fault turnout through the dispatch procedure and the key opening process. Then, the train will run in a mode where the driver or the attendant is responsible for guiding the operation until the route is again recognized by the system with the necessary trust level.
(3) The station communication failed. If the station communication fails, the onboard computer will not be able to obtain the status update information of all signal devices along the track. Therefore, the display of all signal machines in the current control area is regarded as the stop signal. In addition, due to the loss of GPS differential information, onboard computers will start to use speed sensors for accurate positioning. In the same way as the onboard radio failure, the failure of the onboard radio will eventually cause the system to lose its position, so that the system can switch from IN mode to station mode or stay in OUT mode and cannot switch to IN mode.

(4) The trackside VHLC can't receive GPS differential information. Due to the loss of GPS differential information, onboard computers will start to use speed sensors for accurate positioning. In the same way as the onboard radio failure, the failure of the onboard radio will eventually cause the system to lose its position, so that the system can switch from IN mode to station mode or stay in OUT mode and cannot switch to IN mode.

(5) The trackside VHLC can't receive the GPS clock information in time. Timestamp is a method to ensure the correctness of communication data. Therefore, when VHLC still does not receive GPS clock information within 5 min, the system will not be able to judge the timeliness of the information, and the radio will stop sending the information. As a result, the onboard system will lose GPS differential information, and all of the trackside status update data. Then, the onboard computer will deal with the failure according to (6) and (7) proposed in the previous section about the onboard failure analysis.

(6) The RBC database is inconsistent with the onboard computer database. During the train running process, CRC verification and comparison are carried out between the onboard computer and the corresponding database file of RBC. When an RBC database file in a control area is detected to be inconsistent with the corresponding database file of the onboard computer, the onboard computer treats these database files as unusable. The corresponding processing mode is the same as "database validation fails or database file fails" [9].

3.6 Conclusion

This chapter firstly introduces the structure of onboard control unit for ITCS. Then, fault tree analysis method is used to evaluate the reliability of ITCS, so as to obtain the quantitative reliability index of onboard system for ITCS. Furthermore, the corresponding emergency countermeasures and safeguard measures of different fault modes are analyzed.

References

1. Yang HX. Application of GSM-R in modern railway communication. Sci Technol Inf. 2006;28:1–2.
2. GE Global Signaling. Tibet ITCS annex 2-technical specification. 2005.
3. Liao YQ, Li Z. Study and design of ITCS transmission based on GSM-R. Railway Comput Appl. 2006;15(6):42–4.
4. Huang M, Wang CL, Cao JP. Application of ITCS based on GSM-R to Qingzang railway. Railway Comput Appl. 2006;15(7):15–8.
5. Liang H. Analysis on the function and dynamic debugging of ITCS system of Qingzang line. Electric Railway. 2009;3:48–50.

6. Lu ZT. ITCS system of Qingzang line. Railway Signal Commun. 2007;43(8):12–3.
7. Tang JG. Analysis for train control system of Qinghai-Tibet railway based on GPS and GSM-R. J Transp Syst Eng Inf Technol. 2010;10(2):190–4.
8. Qi CJ, Peng DZ. Operational mode analysis of ITCS system on the Qingzang line. Silicon Valley. 2012;7:173.
9. Li K. The project of ITCS signal control system of Qinghai-Tibet railway. Chin Railway. 2005;7:31–6.

Chapter 4
Key Scientific Problems Based on GPS Positioning System

4.1 Overview

In recent years, China's rail transit industry has entered a period of rapid development. The national "12th five-year plan for comprehensive transportation system" has clearly defined the construction plan and high-speed railway construction plan of China's central and Western regions in the future period and also put forward new requirements for railway transportation technology and safety equipment.

As China train control system (CTCS) is widely used in domestic rail transportation, independent train positioning system, including the orientation of wayside balise positioning, onboard sensors positioning, and GPS satellite positioning [1] (Fig. 4.1), is an important part of CTCS-3. Since train positioning is an important part of train operation control system, the level of positioning accuracy has an important impact on the safety and operating efficiency of train, and thus the positioning accuracy has become one of the main technical performance indicators for the train operation control. With the improvement of train operational speed and the further increase of transportation density, the requirements of train operation control system on train positioning accuracy become higher and higher [2]. The importance of train positioning mainly includes the following aspects:

(1) It provides a basis for ensuring the safety interval of the train.
(2) As the main parameter for calculating the speed curve and implementing the braking control, the precise position information of the train is provided for the automatic train operation system (ATO).
(3) As the basic information to display the running state of train, the position information of train is provided for the automatic train supervision (ATS) subsystem.
(4) The position information is used by the automatic train protection (ATP) system for calculating the overspeed protection curve.

D. Chen and R. Cheng, *Intelligent Processing Algorithms and Applications for GPS Positioning Data of Qinghai-Tibet Railway*, https://doi.org/10.1007/978-3-662-58970-0_4

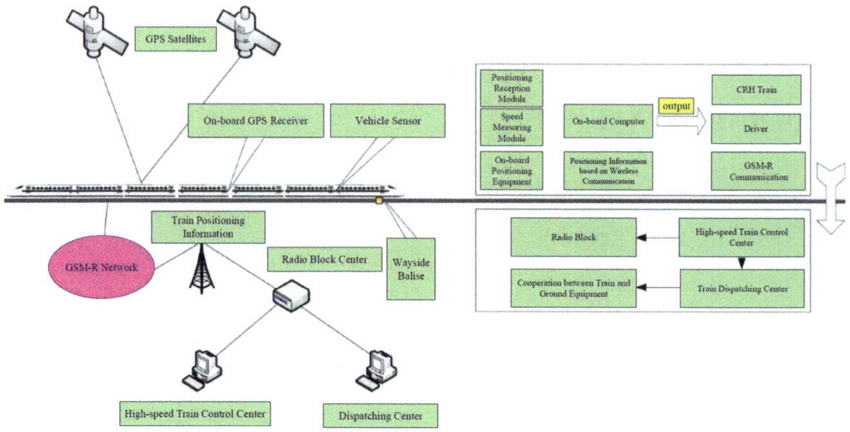

Fig. 4.1 Positioning subsystem of train control system

The user device consists of GPS receiver hardware, data processing software, microprocessor, and terminal equipment. The hardware of GPS receiver includes receiver host, antenna, and power source. Its main function is to receive signals from GPS satellite, obtain necessary navigation information, positioning information, and observation quantity, and realize real-time navigation and positioning of train through simple data processing. GPS software refers to various built-in software, post-processing software, and real-time processing software with differential positioning function or RTK positioning function.

GPS satellite navigation and positioning technology is based on passive ranging principle. As long as the distance between the user's antenna and the GPS satellite is measured by the GPS signal receiver, the position of the user can be calculated.

According to the state of user receiver antenna in measurement, the positioning method based on GPS satellite signal can be divided into static positioning method and dynamic positioning method. It is called static positioning method if the user receiver antenna is stationary during positioning. On the contrary, if the user receiver antenna is in motion during positioning, the adopted positioning method is called dynamic positioning method. For example, dynamic positioning method is used to determine the real-time position of vehicles, ships, and spacecraft.

According to different positions of reference points, it can be divided into absolute positioning and relative positioning method. For absolute positioning method, the absolute position of receiver antenna is determined in the conventional terrestrial system by taking geocentric point as reference point. Because only one receiver is used in the positioning operation, it is also called single point positioning. If a fixed reference point on the ground is selected to determine the position of the relative reference point of the receiver antenna phase center, it is called relative positioning method.

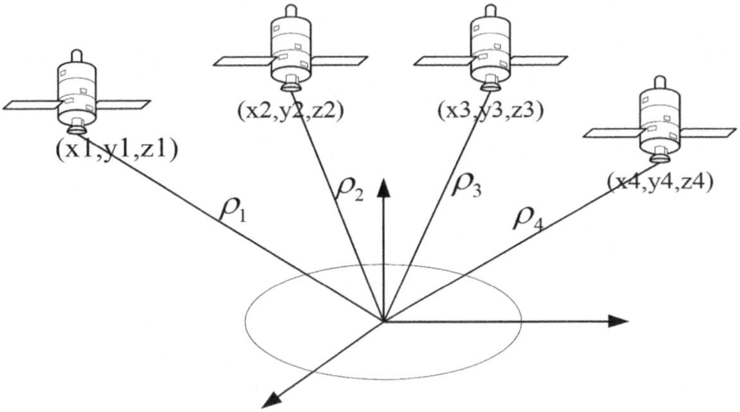

Fig. 4.2 A schematic diagram of pseudo-range positioning principle

Pseudo-range positioning is a common GPS positioning method. GPS satellite can send signals with "pseudo-random noise code" structure according to the spaceborne clock, which is called range code signal (i.e., coarse/acquisition (*C/A*) code or precise (*P*) code). The signal sent from the satellite will reach the receiver antenna after Δt time.

The space geometric distance from the satellite to receiver (ρ) can be obtained by multiplying the signal propagation time Δt by electromagnetic wave speed in a vacuum (c).

$$\rho = \Delta t * c \tag{4.1}$$

In fact, because the propagation time Δt contained the unsynchronized clock error between the satellite clock and the receiver clock, and the propagation delay error of ranging code in the atmosphere, the obtained distance calculated by Eq. (4.1) is not the real geometric distance between the receiving station and the satellite, traditionally called "pseudo range," expressed with ρ. The positioning method corresponding to pseudo-distance is called pseudo-range positioning method. Figure 4.2 shows a schematic diagram of pseudo-range positioning principle.

During the positioning process, each satellite continuously sends a traceable unique coding sequence to the GPS receiver. GPS can identify relevant satellites according to the received coding and then calculate the exact location and accurate time of the receiver. In addition, pseudo-distance method is the basic method of GPS navigation and positioning. The pseudo-distance is calculated by Eq. (4.2).

$$\begin{cases} \rho_r^{-j} = \sqrt{(x^j - x_u)^2 + (y^j - y_u)^2 + (z^j - z_u)^2} - c\left(d_{\text{tr}} - d_{\text{ts}}^j\right) - d_{\text{ion}}^j - d_{\text{trop}}^j \\ \vdots \end{cases}, j = 1, 2, 3, 4, \ldots \tag{4.2}$$

where

ρ_r^{-j}	is the geometric distance between the receiver and the satellite position obtained from the satellite broadcast satellite calendar;
$\{x^j, y^j, z^j\}$	represent the three-dimensional coordinates of the *j*th satellite in the ECEF coordinate system;
d_{tr}	is the receiver clock difference;
d_{ts}^j	is the clock difference of *j*th satellite, $j = 1, 2, \ldots, n$; n is the number of satellites observed simultaneously;
$\{x_u, y_u, z_u\}$	represent the three-dimensional coordinates of the receiver in the ECEF coordinate system;
d_{ion}	is the distance deviation caused by ionospheric effect;
d_{trop}	is the distance deviation caused by tropospheric effect;
C	is the speed of light

The GPS satellite positioning system continuously sends its own calendar parameters and time information through the satellite. After receiving the signal from the GPS, GPS signal receiver can calculate its position according to the triangle formula. Three satellites can realize 2D positioning (longitude, latitude), and four satellites can conduct 3D positioning (longitude, latitude, and height). By constantly updating receiver's information, the movement direction and speed of receiver can be calculated. Since there are 24 GPS navigation satellites in six orbits around the world, there are at least 4 satellites in the horizontal line at any time and at most 11 satellites that can guarantee the positioning accuracy of GPS. The function of GPS signal receiver is used to capture, amplify, and process the satellite signal and calculate the 3D position and speed of the receiver in real time.

4.2 Analysis of Advantages and Disadvantages of Existing Positioning Methods

Speed measurement and positioning provide real-time running speed and specific location information of the train for the train control system. So, its precision directly affects the control precision of the train. The low accuracy of speed measurement and positioning will not only increase the unsafe factors of train operation, but also cause the reserved safety protection distance of the system to be too large, which will affect the transportation efficiency. According to the source of speed information, train speed measurement can be divided into axle speed measurement method and radar speed measurement method. In order to achieve accurate train positioning, which can accurately determine the specific position of the train at any time and at any position, there have been various methods for train positioning so far, which mainly include three categories [3]: (1) train positioning based on track and line-side equipment, e.g., train positioning method based on track circuit or balise; (2) train positioning based on sensors, e.g., train positioning based on odometer, radar speed measurement, track electronic map, and inertial navigation system; (3) train positioning based on satellite navigation [4], e.g., GPS satellite positioning and so on. For the train control system, the single positioning method has limitations in practical application. On July 23, 2011, for example, the serious railway traffic accident of Yong Wen line is due to the failure of track circuit caused by a lightning strike. The train control center has lost the positioning information of the train in front of track circuit and still controls the railway signals according to the condition that no train occupied in the section. Then, the wrong control state of interval signal machine directly or indirectly caused the accident. Therefore, the high-speed train positioning generally adopts multi-sensor fusion positioning or GPS/INS/MM combined positioning system since then [5–9]. The multi-sensor integrated positioning system uses the axle speed sensor, doppler radar, and accelerometer to measure the real-time speed and acceleration of the train. The multi-stage Kalman filtering algorithm or $H\infty$ filtering algorithm can be used as fusion algorithm to process the information fusion, eliminate the measurement error and random error system. The advantage of this method is that through independent measurement of multiple sensors, it can reduce information error, information redundancy and supplement each other, so as to improve the reliability of the measurement results. At the same time, when the speed measurement system fails, a better fault-tolerant algorithm can guarantee a certain accuracy, so as not to bring the breakdown of the whole system. However, this method relies heavily on infusion algorithms. When the sensor information is lost, it is difficult to measure 3d coordinates, which will result in the increase of measurement error, the decrease of system precision. The GPS/INS/MM integrated positioning system uses the rapid difference of the network-based real-time kinematic (RTK) technology to reduce the positioning error and modify the information measured by GPS, such as train position, speed, and acceleration. By relying on the results of GPS measurement, this method can eliminate the measurement error of inertial unit in time, which can make maneuverability flexible and adaptation to environment well.

Table 4.1 shows the comparison of train positioning methods. In Table 4.1, PT, CP, PA, ICs, MCs, MFs, MAs, and MDs mean positioning technology, continuous positioning, positioning accuracy, investment costs, maintenance costs, maneuvering features, main advantages, major defects, respectively.

4.2.1 Train Positioning Based on Query Balise

Query balise is a kind of high-speed point data transmission equipment based on electromagnetic coupling principle. As a key component of the ATP system, query balise is used to exchange data between the train and ground equipment in a specific location. It can provide the needed point of information for ATP to ensure the safety of train operation at high-speed condition, including the route length, length of turnout area, block section length, slope, curve, etc. In addition, the query

Table 4.1 Comparison of train positioning methods

PT	PP	CP	ICs	MCs	MFs	MAs	MDs
Track circuit	No	Poor	High	High	Poor	Simple, mature technology, safe and reliable	Small transmission information, not continuous positioning, poor mobility, large maintenance
Balise	No	High	High	High	Poor	The transmission information amount is large, it can provide the absolute position information, the auxiliary localization performance is good	Only point information can be given, large quantity of balise is required, high cost of investment and maintenance
Axle counter-positioning	No	Poor	High	High	Poor	Simple and reliable, strong adaptability to the environment and good performance of auxiliary positioning	Poor positioning accuracy and high maintenance costs
Odometer	Yes	Higher	Higher	Low	Good	Simple, high measurement accuracy for high-speed, less environmental impact	Depending on the wheel rotation, slip, idle, which can cause accumulated error
Doppler radar	Yes	Higher	Low	Low	Good	Independent positioning, not affected by the wheel, wide adaptability, mature technology	Installation technology requires high requirements, accuracy will affect by the accumulated error caused by body vibration, ground bump, installation angle
Satellite [10] positioning	Yes	Low	High	High	Good	It can be used both as a means of communication and as a means of measuring train position	A large number of trackside equipment is needed, and it has signal blind area and the reliability needs to be improved

balise includes ground equipment and onboard equipment. Ground equipment mainly includes ground balise. Onboard equipment mainly includes onboard query antenna and onboard query host. Speed sensors can be used to measure the relative travel distance of the train, but if the initial position (or reference position) of the train is unknown, the sensors cannot be used to determine the absolute position direction relative to the earth. When the train receives the fixed position of the train by balise, the accumulated error will be eliminated. On the basis of the speed sensor, the train positioning accuracy will be improved by adding beacon at the appropriate location on the ground.

The line-side query balise can provide accurate train position information to ensure the normal operation of train control system. However, in the Qinghai–Tibet railway, the environment is complex, the altitude is high, and the population is small. If a lot of balises are arranged, it is difficult to carry out long-term and continuous maintenance work. Therefore, the positioning method is not only dependent on the query balise in the Qinghai–Tibet railway.

4.2.2 Intelligent Localization Method by Using LS-SVM and the Online Sparse Optimization Approach

For train control system, the train-ground communication is achieved through bidirectional wireless communication, which means that onboard unit (OBU) and radio block center (RBC) can communicate with each other through Global System for Mobile Communication Railway (GSM-R) networks. Figure 4.4 illustrates the bidirectional wireless communication process in HST control system. The main function of RBC is to allocate movement authority (MA) for high-speed trains (HSTs) in its jurisdiction zones. After receiving the wayside equipment information from RBC by GSM-R, OBU will generate dynamic train speed profile (DTSP), which includes service brake intervention curve (SBI), emergency brake intervention curve (EBI), and released intervention curve (RELI). Then, the HST will be controlled by DTSP.

As shown in Fig. 4.3, the horizontal axis shows the HST's location, and the vertical axis shows the speed of the following HST. Safety distance (SD) is the supplementary distance of service brake's distance to ensure the HST's safety in the worst condition. The End of MA for SB (A) and the End of MA for EB (B) represent the terminal point of SBI and EBI, respectively. Start Braking Distance (SBD) is the location point where the HST starts braking. As the precise control of the train is realized on the basis of accurate localization, the train localization is of great significance to ensure the safety of the train in the moving block control system (MBCS). In addition, the train localization has a great influence on the operation efficiency of railway and HSTs since the location accuracy affects the length of SD.

For the current localization system, the absolute location information of HST is checked and corrected by the balise equipment and the relative location of HST between two adjacent balises is calculated by pulse speed sensor (PSS). The location

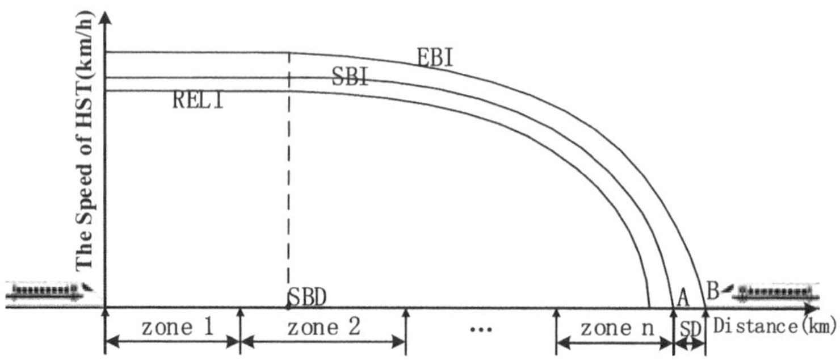

Fig. 4.3 DTSP of the following HST in moving block control system

error of PSS includes two inevitable factors: counting error and wheel diameter loss. In addition, counting error and wheel diameter loss are mainly caused by idling, sliding and creep of wheels and wheel abrasion, respectively. The location error of PSS will lead to the occurrence of accumulated location error between two adjacent balises. For the traditional quasi-MBCS, the headway of two adjacent HSTs is related with the block section division. In other words, following train can't run in the same block section with the leading train and the following train must stop running before the boundary of block section occupied by the leading train. However, more than two adjacent HSTs can run in the same block section in MBCS. Under this circumstance, the location correction of balises makes no sense, and the accumulated location error of PSS between two adjacent balises will endanger the safety of HSTs.

As shown in Figs. 4.4 and 4.5, when the train passes over the balise, the absolute position information of the set point sent by balise or track will update the train position calculated through laying a balise or track loop at every certain distance along the railway. Then, a new starting point of train positioning will be obtained and the cumulative positioning error measured by the sensor will be eliminated. In addition, for the detailed information about the intelligent train positioning model, we can refer to [11–13].

The working principle of balises is shown in Fig. 4.6. From Fig. 4.6, we can see that the cumulative positioning error (CPE) between two adjacent balises (B_i and B_{i+1}) accumulates until the next adjacent balise B_{i+1}.

Based on the position report created by the onboard equipment, the basic flowchart of intelligent localization method of HST is depicted in Fig. 4.7. In order to improve the real-time performance and location accuracy of HST, the proposed method comprises two parts, including off-line modeling process and online updating process. Because the operation of HST can produce a lot of historical location data, the location data of HST belongs to large-scale data. So, we use off-line sparse algorithms to process the historical data in the first part, such as LS-SVM-L_0 NM, LS-SVM-L_0 NM, and LS-SVM-IPEM. Then, the localization model is obtained by training the sparse off-line historical data.

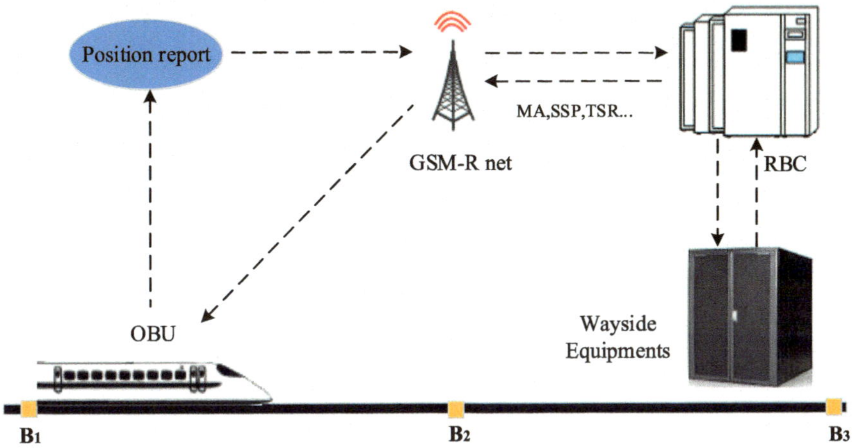

Fig. 4.4 Train positioning based on balise

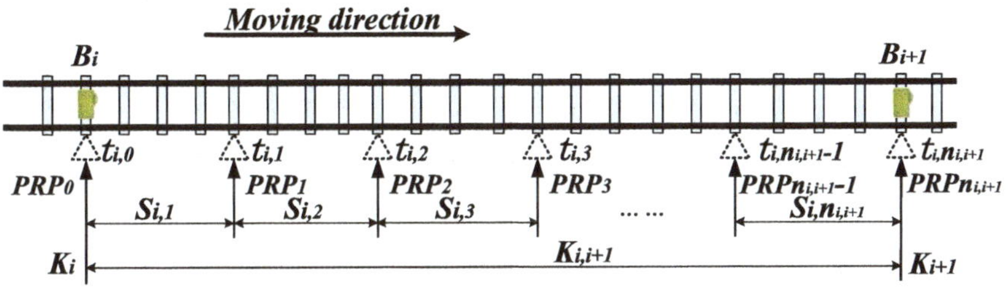

Fig. 4.5 Intelligent train positioning model based on balise

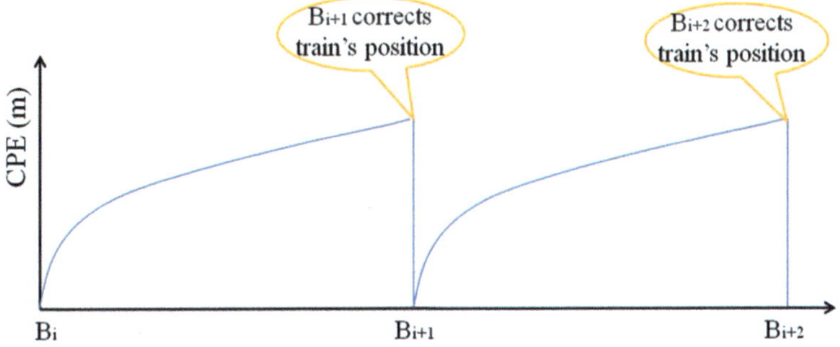

Fig. 4.6 Working principle of balises

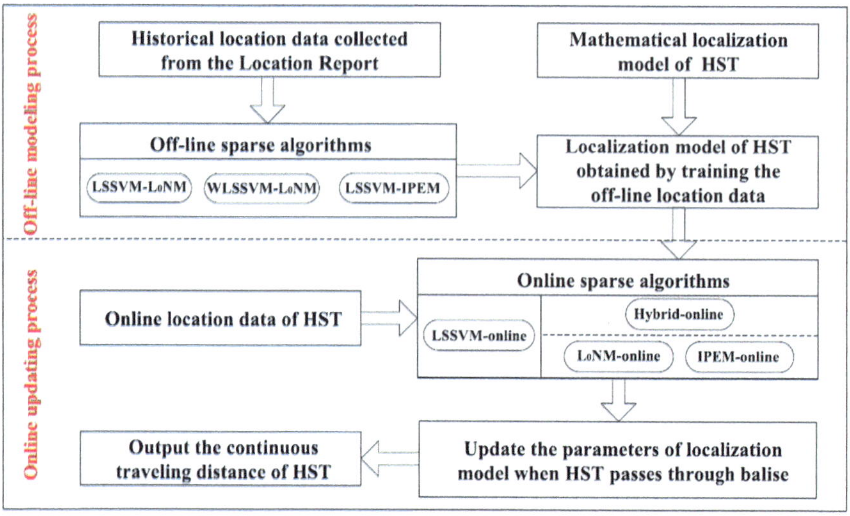

Fig. 4.7 Flowchart of intelligent localization method

Different from subway train, the HST speed is higher than that of subway train and the operational time interval between two adjacent railway stations is longer than that of subway train. For this reason, HST also produces large amounts of location data online. To ensure the online real-time performance of localization method and improve the adaptability, the new location data is added into the training data set and sparsified by online sparse algorithms, such as LS-SVM-online, NM-online, IPEN-online, and Hybrid-online. Parameters of HST's localization model will be updated when HST passes through balise every time and thus the continuous traveling distance of HST can be estimated online.

In order to establish the localization model, we assume that there are $n_{i,i+1} + 1$ location report points (LRPs) between adjacent balises, which includes the LRPs on the two adjacent balises. $n_{i,i+1}$ means the number of location reports that HST sends between two adjacent balises. The mathematical localization model is shown in Fig. 4.5. B_i and B_{i+1} represent the adjacent balises; K_i and K_{i+1} are the kilometer mark corresponding to B_i and B_{i+1}, respectively; $t_{i,j}$ represents the operational time on the jth LRP between B_i and B_{i+1}; $S_{i,j}$ represents the traveling distance from the $(j-1)$th LRPs to the jth LRPs between B_i and B_{i+1}; $K_{i,i+1}$ is the actual distance between adjacent balises B_i and B_{i+1}, and it can be calculated by the kilometer mark of the adjacent balises.

$$K_{i,i+1} = K_{i+1} - K_i \tag{4.3}$$

The train's location is determined by the kilometer mark and distance from the recently related balises. Thus, we can obtain any LRPs' location between the adjacent balises from Fig. 4.5.

$$L_{i,j} = K_i + \sum_{m=1}^{j} S_{i,m},\ j = 1, 2, \ldots, n_{i,i+1} \tag{4.4}$$

According to the CTCS technical standard, the OBU should send the location report information of train to RBC about every 7 s. Therefore, we allocate the error to each distance between adjacent LRPs to estimate the accurate traveling distance $S^*_{i,m}$ between the adjacent LRPs. The specific computation steps are summarized as follows:

Step 1. Take the same balise B_i as the recently related balise to obtain the train running data, which includes the speed and the time on LRPs $(V_{i,j}, t_{i,j})$ and the kilometer mark of the recently related balise (K_i).
Step 2. Obtain the distance $S^a_{i,j}$ between any adjacent LRPs that is measured by pulse speed sensors.
Step 3. Sum all the distance $S^a_{i,j}$ between adjacent LRPs to obtain the estimated distance between the balise B_i and B_{i+1}.

$$K^a_{i,i+1} = \sum_{j=1}^{n_{i,i+1}} S^a_{i,j} \tag{4.5}$$

Step 4. Calculate the error $e^a_{i,i+1}$ between the adjacent balise B_i and B_{i+1}.

$$e^a_{i,i+1} = K_{i,i+1} - K^a_{i,i+1} \tag{4.6}$$

Step 5. Allocate the error $e^a_{i,i+1}$ to each part $S^a_{i,j}$ for estimating the accurate traveling distance between adjacent LRPs.

$$S^*_{i,j} = S^a_{i,j} + S^a_{i,j} * \frac{e^a_{i,i+1}}{K^a_{i,i+1}},\quad j = 1, 2, \ldots, n_{i,i+1} \tag{4.7}$$

Based on the above assumption, the train localization problem is converted into a regression problem. In other words, we need to find a regression function with $V_{i,j-1}, V_{i,j}, \Delta t_{i,j-1,j}$ as inputs, and its output can be as closely as $S^*_{i,j}$. So, we use LS-SVM as the regression function and mainly concentrate on designing the sparse optimization algorithms to reduce the SVs number of LS-SVM model.

In this following, we propose three off-line sparse algorithms for train localization. Firstly, we introduce the localization model based on LS-SVM. Then, three off-line sparse algorithms, such as LS-SVM-L_0 NM, WLS-SVM-L_0 NM and LS-SVM-IPEM, are proposed by using L_0-norm minimization and pruning error minimization algorithms, respectively.

A. Least Squares Support Vector Machines

For solving of a linear system instead of a QP problem, LS-SVM has good capability of generalization and less complexity than SVM. Therefore, we propose a localization model to obtain the continuous location information between two adjacent LRPs by using LS-SVM model. The location model based on LS-SVM is computed by Eq. (4.8).

$$S_{i,j}^{\text{ls-svm}} = f\left(V_{i,j-1}, V_{i,j}, \Delta t_{i,j-1,j}\right) = \boldsymbol{\omega}^T \varphi\left(x_{i,j-1,j}\right) + b \tag{4.8}$$

where $x_{i,j-1,j}$ represents input vector $\left[V_{i,j-1}, V_{i,j}, \Delta t_{i,j-1,j}\right]^{\mathrm{T}}$. $\varphi\left(x_{i,j-1,j}\right)$ maps $x_{i,j-1,j}$ to a vector in a feature space M. And $\boldsymbol{\omega}$ is a vector in M. The $\boldsymbol{\omega}$ and b in Eq. (4.8) are obtained by solving the following optimization problem:

$$\min_{\boldsymbol{\omega},b,\boldsymbol{e}} J(\boldsymbol{\omega}, \boldsymbol{e}) = \frac{1}{2}\boldsymbol{\omega}^T \boldsymbol{\omega} + \frac{1}{2}\gamma \sum_{k=1}^{N} e_k^2 \tag{4.9.a}$$

$$s.t.\ y_k = \boldsymbol{\omega}^T \varphi(x_k) + b + e_k, \quad \forall k = 1, 2, \ldots, N. \tag{4.9.b}$$

The Lagrangian function is given by

$$L(\boldsymbol{\omega}, b, \boldsymbol{e}, \boldsymbol{\alpha}) = J(\boldsymbol{\omega}, \boldsymbol{e}) - \sum_{k=1}^{N} \alpha_k \left\{\boldsymbol{\omega}^T \varphi(\boldsymbol{x}_k) + b + e_k - y_k\right\} \tag{4.10}$$

where $\alpha_k \geq 0$ is Lagrangian multipliers.

The conditions for optimality are as follows:

$$\begin{cases} \frac{\partial L}{\partial \boldsymbol{\omega}} = 0 \Rightarrow \boldsymbol{\omega} = \sum\limits_{k=1}^{N} \alpha_k \varphi(\boldsymbol{x}_k) \\ \frac{\partial L}{\partial b} = 0 \Rightarrow \sum\limits_{k=1}^{N} \alpha_k = 0 \\ \frac{\partial L}{\partial e_k} = 0 \Rightarrow \alpha_k = \gamma e_k \\ \frac{\partial L}{\partial \alpha_k} = 0 \Rightarrow \boldsymbol{\omega}^T \varphi(\boldsymbol{x}_k) + b + e_k - y_k = 0 \end{cases} \tag{4.11}$$

for $k = 1, 2, \ldots, N$. After elimination of $\boldsymbol{\omega}$ and e_k, the solution can be represented by the following set of linear equations:

$$\begin{pmatrix} 0 & \mathbf{1}_n^T \\ \mathbf{1}_n & \Omega + \gamma^{-1}\boldsymbol{I} \end{pmatrix} \begin{pmatrix} b \\ \boldsymbol{\alpha} \end{pmatrix} = \begin{pmatrix} 0 \\ \boldsymbol{y} \end{pmatrix} \tag{4.12}$$

with $\boldsymbol{y} = (y_1, y_2, \ldots, y_n)^{\mathrm{T}}$, $\mathbf{1}_n = (1, 1, \ldots, 1)^{\mathrm{T}}$, $\boldsymbol{\alpha} = (\alpha_1, \alpha_2, \ldots, \alpha_n)^{\mathrm{T}}$, and $\Omega_{kl} = K(\boldsymbol{x}_k, \boldsymbol{x}_l) = \varphi(\boldsymbol{x}_k)^T \varphi(\boldsymbol{x}_l)$ for $k = 1, 2, \ldots, N$. We focus on the choice of an RBF: $K(\boldsymbol{x}_k, \boldsymbol{x}_l) = \exp\left(-\frac{\|x_k - x_l\|^2}{2\sigma^2}\right)$ for the sequel and the variance of the Gaussian kernel $\sigma > 0$. According to Mercer's theorem, the location estimation function of the resulting LS-SVM model becomes $f(\boldsymbol{x}) = \sum_{k=1}^{N} \alpha_k K(\boldsymbol{x}_k, \boldsymbol{x}_l) + b$ where $K(\cdot, \cdot)$ is an appropriately chosen positive definite kernel.

B. Sparse LS-SVM Model Based on L_0-Norm Minimization Algorithm

In order to improve the sparsity of the LS-SVM model, we transform the LS-SVM model described by Eqs. (4.9.a) and (4.9.b) into the following primal problem:

$$\min_{\boldsymbol{\alpha}, b, e_k} \frac{1}{2} \sum_{k=1}^{N} \lambda_k \alpha_k^2 + \frac{1}{2} \gamma \sum_{k=1}^{N} e_k^2 \tag{4.13.a}$$

$$\text{s.t.} \sum_{l=1}^{N} \alpha_l \boldsymbol{\Omega}_{kl} + b = y_k - e_k, \quad \forall k = 1, 2, \ldots, N. \tag{4.13.b}$$

Compared with the LS-SVM model described by Eqs. (4.9.a) and (4.9.b), the sparse LS-SVM with L_0-norm minimization expressed by Eqs. (4.13.a) and (4.13.b) don't contain an explicit $\boldsymbol{\omega}\boldsymbol{\psi}$ vector, which is denoted by an implicit feature map $\boldsymbol{\omega} = \sum_{l=1}^{N} \alpha_l \Phi(X_l) \boldsymbol{\psi}$. The regularization term is not on $\leftarrow \|\boldsymbol{\omega}^2\|$, but on the $\|\boldsymbol{\alpha}^2\|$ vector, via the prefixed coefficients λ_k. By introducing coefficients β for the Lagrangian function and combing the conditions for optimality, we can formulate the matrix Eq. (4.14):

$$\begin{pmatrix} 0 & \mathbf{1}_n^T \\ \mathbf{1}_n & \boldsymbol{H} \end{pmatrix} \begin{pmatrix} b \\ \boldsymbol{\beta} \end{pmatrix} = \begin{pmatrix} 0 \\ \boldsymbol{y} \end{pmatrix} \tag{4.14}$$

with $\boldsymbol{H} = \boldsymbol{\Omega}\mathrm{diag}(\lambda)^{-1}\boldsymbol{\Omega} + \gamma^{-1}\boldsymbol{I}$, $\boldsymbol{\Omega}$ is the kernel matrix, and γ is regularization parameter. For solving this optimization problem, the sparsity procedures consist of solving iteratively system (4.14) for different values of $\boldsymbol{\lambda}$. In **Algorithm 1**, APE_{AB} represents the absolute percentage error between adjacent balises and $\boldsymbol{K}$ is the kernel matrix.

Algorithm 1 Off-line LSSVM sparse algorithm based on L_0-norm minimization (LSSVM-L_0NM).

Given existing training data $\{x_k, y_k\}, k = 1, 2, \cdots, N$, solve linear systems (4.12) to obtain optimal (γ, σ^2) and $(\boldsymbol{\alpha}, b)$ by 10-fold cross-validation.
$\lambda_i \leftarrow \alpha_i, i = 1, 2, \cdots, N$.
repeat
    $\boldsymbol{H}^t \leftarrow \boldsymbol{K}\mathrm{diag}(\boldsymbol{\lambda}^t)^{-1}\boldsymbol{K} + \boldsymbol{I}_N/\gamma, i = 1, 2, \cdots, N$.
    Solve system (4.14) to obtain $\boldsymbol{\beta}$ and $\bar{b}$.
    $\bar{\boldsymbol{\alpha}}^t \leftarrow \mathrm{diag}(\boldsymbol{\lambda}^t)^{-1}\boldsymbol{K}\boldsymbol{\beta}$.
    if $|\bar{\alpha}_i^t| > \epsilon$ **then**
        $\lambda_i^{t+1} \leftarrow 1/(\bar{\alpha}_i^t)^2, i = 1, 2, \cdots, N$.
    else
        $\lambda_i^{t+1} \leftarrow 1/\epsilon^2, i = 1, 2, \cdots, N$.
    end if
until convergence
Delete the corresponding position data $\{x_i, y_i\}$ when $|\bar{\alpha}_i^*| < 10^{-6}$ and update N to obtain the new training data set $\{sx_k, sy_k\}, k = 1, 2, \cdots, N$.
Given training data $\{sx_k, sy_k\}, k = 1, 2, \cdots, N$, solve linear systems (4.12) to obtain γ^*, σ^{2*}, $\boldsymbol{\alpha}^*$ and b^*.
Compute the transient traveling distance by $y(\boldsymbol{x}) = \sum_{k=1}^{N} \alpha_k^* \boldsymbol{K}(\boldsymbol{x}, \boldsymbol{x}_k) + b^*$ and APE_{AB} value.
return APE_{AB}.

C. Sparse-Weighted LS-SVM Model Based on L_0-Norm Minimization Algorithm

In order to improve the sparsity of the LS-SVM model, we transform the LS-SVM model described by Eqs. (4.9.a) and (4.9.b) into the following primal problem:

$$\min_{\boldsymbol{\alpha}, b, e_k} \frac{1}{2}\sum_{k=1}^{N} \lambda_k \alpha_k^2 + \frac{1}{2}\gamma \sum_{k=1}^{N} v_k e_k^2 \tag{4.15.a}$$

$$\text{s.t.} \sum_{l=1}^{N} \alpha_l \boldsymbol{\Omega}_{kl} + b = y_k - e_k, \quad \forall k = 1, 2, \ldots, N. \tag{4.15.b}$$

Introducing coefficients β, we define the Lagrangian function to be the following equation:

$$\boldsymbol{L}(\boldsymbol{\alpha}, b, \boldsymbol{e}, \boldsymbol{\beta}) = \frac{1}{2}\sum_{k=1}^{N} \lambda_k \alpha_k^2 + \frac{1}{2}\gamma \sum_{k=1}^{N} v_k e_k^2 - \sum_{k=1}^{N} \beta_k \left\{ \sum_{l=1}^{N} \alpha_l \Omega_{kl} + b + e_k - y_k \right\} + \alpha_j^m K\left(x_j, x_j\right) + b^m - b^{m+1} \tag{4.16}$$

where $\beta_k \geq 0$ is Lagrangian multipliers.

By combing the conditions for optimality, we can formulate the matrix Eq. (4.17):

$$\begin{pmatrix} 0 & \mathbf{1}_n^T \\ \mathbf{1}_n & \boldsymbol{L} \end{pmatrix} \begin{pmatrix} b \\ \boldsymbol{\beta} \end{pmatrix} = \begin{pmatrix} 0 \\ \mathbf{y} \end{pmatrix} \tag{4.17}$$

with $\boldsymbol{L} = \boldsymbol{\Omega}\mathrm{diag}(\lambda)^{-1}\boldsymbol{\Omega} + V_\gamma$, and $V_\gamma = \mathrm{diag}\left\{\frac{1}{\gamma v_1}, \frac{1}{\gamma v_2}, \ldots, \frac{1}{\gamma v_n}\right\}$. The weights v_k are determined by the error variables $e_k = \alpha_k/\gamma$ from the LS-SVM case (4.9) and are computed by Eq. (4.18) [14].

$$v_k = \begin{cases} 1 & \text{if } \left|e_k/\hat{s}\right| \le c_1 \\ \frac{c_2 - \left|e_k/\hat{s}\right|}{c_2 - c_1} & \text{if } c_1 \le \left|e_k/\hat{s}\right| \le c_2 \\ 10^{-4} & \text{otherwise} \end{cases} \tag{4.18}$$

where $\hat{s}$ is a robust estimation of the standard deviation of the LS-SVM error variable e_k and is computed by $\hat{s} = 1.483 * \mathrm{MAD}(x_i)$. MAD represents the median absolute deviation of the error variables e_k. Considering the distribution of error e_k, we describe the weighted LS-SVM sparse algorithm based on L_0-norm minimization in **Algorithm 2**.

Algorithm 2 Weighted LSSVM sparse algorithm based on L_0-norm minimization (WLSSVM-L_0NM).

1: Given training data $\{x_k, y_k\}, k = 1, 2, \cdots, N$, solve linear systems (4.12) to obtain optimal (γ, σ^2) and ($\boldsymbol{\alpha}$, b) by 10-fold cross-validation.
2: Compute error by $e_k = \alpha_k/\gamma$ based on the optimal γ and α.
3: Compute $\hat{s}$ from the e_k distribution.
4: Determine the weights v_k based upon e_k, $\hat{s}$ by Eq.(4.18).
5: $\lambda_i \leftarrow \alpha_i, i = 1, 2, \cdots, N$.
6: **repeat**
7:     $\boldsymbol{L}^t \leftarrow \boldsymbol{K}\mathrm{diag}(\boldsymbol{\lambda}^t)^{-1}\boldsymbol{K} + \underline{\mathrm{V}}_\gamma, i = 1, 2, \cdots, N$.
8:     Solve system (4.17) to obtain $\boldsymbol{\beta}$ and $\bar{b}$.
9:     $\bar{\boldsymbol{\alpha}}^t \leftarrow \mathrm{diag}(\boldsymbol{\lambda}^t)^{-1}\boldsymbol{K}\boldsymbol{\beta}$.
10:     **if** $|\bar{\alpha}_i^t| > \epsilon$ **then**
11:         $\lambda_i^{t+1} \leftarrow 1/(\bar{\alpha}_i^t)^2, i = 1, 2, \cdots, N$.
12:     **else**
13:         $\lambda_i^{t+1} \leftarrow 1/\epsilon^2, i = 1, 2, \cdots, N$.
14:     **end if**
15: **until** convergence
16: Execute steps 13-16 of ***Algorithm*** 1.

D. Sparse LS-SVM Model Based on Iterative Pruning Error Minimization Algorithm

Different from LS-SVM-L_0 NM and WLS-SVM-L_0 NM, the introduced error is considered in pruning error minimization algorithm (PEM) [15]. And, only the sample that bears the smallest error after it is omitted will be deleted from the original data set. The evaluation of the output at iteration m of training sample x_j can be computed by Eq. (4.19).

$$\hat{y}^m(x_j) = \sum_{\substack{k=0 \\ k \neq j}}^{N} \alpha_k^m K(x_k, x_j) + \alpha_j^m K(x_j, x_j) + b^m \tag{4.19}$$

where α_k^m and b^m denote the multiplier for sample k and the bias at iteration m. If sample j is removed from the training set, the output of x_j in iteration $m+1$ will be calculated by Eq. (4.20). So, the introduced error of sample j described by $d(x_j)$ is computed by Eq. (4.21) when sample x_j is omitted from the training data set.

$$\hat{y}^{m+1}(x_j) = \sum_{\substack{k=0 \\ k \neq j}}^{N} \alpha_k^{m+1} K(x_k, x_j) + b^{m+1} \tag{4.20}$$

$$d(x_j) = \hat{y}^m(x_j) - \hat{y}^{m+1}(x_j) = \sum_{\substack{k=0 \\ k \neq j}}^{N} \left(\left(\alpha_k^m - \alpha_k^{m+1} \right) K(x_k, x_j) \right) + \alpha_j^m K(x_j, x_j) + b^m - b^{m+1} \tag{4.21}$$

As shown in Eq. (4.21), the introduced error does not just depend on α_j^m, but more other parameters have influence on the introduced error. The regularization parameter γ is used to weight the importance of accurate approximation and the smoothness of the function. When the γ is set to zero, the approximation error does not influence the $J(\boldsymbol{\omega}, \boldsymbol{e})$, so only the weight $\boldsymbol{\omega}$ is minimized. When $\gamma = \infty$, System (4.12) can be converted to (4.22). That is $\boldsymbol{Ax} = \boldsymbol{c}$, in which $\boldsymbol{A}$ is the matrix on the left-hand side, $\boldsymbol{x}$ is the Lagrangian multiplier $\boldsymbol{\alpha}$ and the bias b, $\boldsymbol{c}$ is the target vector $(0, \mathbf{y})^{\mathrm{T}}$.

$$\begin{pmatrix} 0 & \mathbf{1}_n^T \\ \mathbf{1}_n & \Omega \end{pmatrix} \begin{pmatrix} b \\ \boldsymbol{\alpha} \end{pmatrix} = \begin{pmatrix} 0 \\ \mathbf{y} \end{pmatrix} \tag{4.22}$$

In the case of $\gamma \neq \infty$, the introduced error can be obtained by a similar derivation process. Instead of the un-regularized matrix $\boldsymbol{A}$, the regularized matrix $\boldsymbol{A}_\gamma$ is calculated by Eq. (4.23). And the smallest absolute value of omitting the jth sample can be calculated by Eq. (4.24). The detailed verification process can refer to [15].

$$\boldsymbol{A}_\gamma = \boldsymbol{A} + \gamma^{-1} \boldsymbol{I}_{(n+1)} \tag{4.23}$$

$$d(x_j) = \left[\frac{\boldsymbol{A}\boldsymbol{A}_\gamma^{-1} \boldsymbol{e}_j \boldsymbol{e}_j^T \boldsymbol{A}_\gamma^{-1}}{\boldsymbol{e}_j^T \boldsymbol{A}_\gamma^{-1} \boldsymbol{e}_j} \boldsymbol{c} \right]_j \tag{4.24}$$

where $\boldsymbol{e}_j$ is a column vector of size $N+1$ filled with zeros except element $j+1$ which is equal to one. The omission of a sample will introduce an error at other samples on account of the regularization parameter γ is not infinite anymore. In this part, we propose iterative pruning error minimization (IPEM) to obtain sparse LS-SVM model. The detailed procedures are described in **Algorithm 3**.

Algorithm 3 Off-line LSSVM sparse algorithm based on iterative pruning error minimization (LSSVM-IPEM).

1: Given training data $\{x_k, y_k\}, k = 1, 2, \cdots, N$, solve linear systems (4.12) to obtain optimal (γ, σ^2) and ($\boldsymbol{\alpha}$, b) by 10-fold cross-validation.
2: Compute the matrix $\boldsymbol{A}$ and $\boldsymbol{A}_\gamma$ by Eq.(4.22) and Eq.(4.23).
3: Compute the introduced error $d(x_j), j = 1, 2, \cdots, N$. by Eq.(4.24).
4: Sort the values $|d(x_j)|, j = 1, 2, \cdots, N$.
5: Remove a small amount of M points ($\lambda = 100 * M/N, 0 < \lambda \leq 0.9$) that have the smallest introduced errors in the sorted $|d(x_j)|$ spectrum.
6: Retain $N - M$ points and set $N := N - M$.
7: Go to step 1 and retrain on the reduced training data set $\{x_k, y_k\}, k = 1, 2, \cdots, N$, unless the performance of APE_{AB} degrades.

As the HST's running environment changes constantly, to increase the adaptive ability and real-time response of proposed localization model, online sparse methods are developed to update the localization model in the HST's operation process. On the one hand, for increasing the adaptive ability of the LS-SVM model, new location data will be added to the training data set when HST passes through balise. However, retraining the LS-SVM model is time-consuming when the training data set is very large. And, the new added location samples are not always better than the old ones. On the other hand, to enhance the real-time performance of proposed localization model, the training data set must be sparsified dynamically by using sparse algorithms online on the premise of reducing the location error.

Based on the analysis of off-line sparse models, we design the online sparse models using L_0-norm minimization and iterative pruning error minimization algorithms to meet the real-time requirements of HST control system. There are four sparse models, such as online sparse LS-SVM model based on basic pruning algorithm (*LS-SVM-online*), online sparse LS-SVM model based on L_0-norm minimization algorithm (L_0 *NM-online*), online sparse LS-SVM model based on iterative pruning error minimization algorithm (*IPEM-online*), and online sparse LS-SVM model based on Hybrid algorithm (*Hybrid-online*), which combines the L_0 *NM-online* and *IPEM-online*. During the update process, the objective is to select important location data by deleting the training data with little contribution to the location estimation.

Based on the basic pruning algorithm, we design *LS-SVM-online*. The main procedures of *LS-SVM-online* are shown in **Algorithm 4**. Because this pruning algorithm has little sparseness ability, the sparse coefficient λ must be no more than 0.1 on the premise of reducing the location error.

Algorithm 4 LSSVM-online.

Input:

The existing data set, $\{x_k, y_k\}, k = 1, 2, \cdots, N$;
The balises' number on the railway line, Num_{bal};

Output:

HST's location, the sparsified data set: $\{sx_k^i, sy_k^i\}, k = 1, 2, \cdots, N$;

1: **for** i=1 to Num_{bal} **do**
2: Given training data $\{x_k^i, y_k^i\}, k = 1, 2, \cdots, N$, solve linear system (4.12) to obtain optimal parameters (γ^i, σ^{2i}) and ($\boldsymbol{\alpha}^i$, b^i) by 10-fold cross-validation.
3: Sort the values $|\alpha_k^i|, k = 1, 2, \cdots, N$.
4: Remove a small amount of M points ($\lambda^i = 100 * M/N, 0 < \lambda^i \leq 0.1$) that have the smallest values in the sorted $|\alpha_k^i|$ spectrum.
5: Retain $N - M$ points and set $N := N - M$.
6: Go to step 2 and retrain the sparsified training set $\{x_k^i, y_k^i\}, k = 1, 2, \cdots, N$, unless the performance of APE_{AB}^i degrades.
7: Add the latest location report data between adjacent balises in $\{x_k^i, y_k^i\}, k = 1, 2, \cdots, N$, and update N to form new training data set $\{x_k^{i+1}, y_k^{i+1}\}, k = 1, 2, \cdots, N$.
8: **end for**
9: $\{sx_k^i, sy_k^i\} \leftarrow \{x_k^i, y_k^i\}, k = 1, 2, \cdots, N$.

Based on L_0-norm minimization algorithm, we design L_0-norm-online. The main procedures of L_0-norm-online are shown in **Algorithm 5**. When APE_{AB}^{i} is bigger than 1%, it is clear that the current LS-SVM model can't adapt to the change of environment or HST. So, we add the new location data in the training data set and update the LS-SVM model. Otherwise, we sparsify the training data set by using **Algorithm 1** and then model the LS-SVM model with the sparsified training data set.

Algorithm 5 L_0NM-online.

Input:

The sparsified data set by using LSSVM-L_0NM algorithm, $\{x_k, y_k\}, k = 1, 2, \cdots, N$;
The balises' number on the railway line, Num_{bal};

Output:

HST's location, the sparsified training data set: $\{sx_k^i, sy_k^i\}, k = 1, 2, \cdots, N$;

1: **for** i=1 to Num_{bal} **do**
2: Given training data $\{x_k^i, y_k^i\}, k = 1, 2, \cdots, N$, solve linear system (4.12) to obtain optimal parameters (γ^i, σ^{2i}) and ($\boldsymbol{\alpha}^i$, b^i) by 10-fold cross-validation.
3: Predict the HST's location by the established LSSVM model in the previous step and compute APE_{AB}^i.
4: **if** $APE_{AB}^i \geq 1\%$ **then**
5: Execute step 7 of ***Algorithm*** 4.
6: **else**
7: Sparsify the training data set $\{x_k^i, y_k^i\}, k = 1, 2, \cdots, N$ by using ***Algorithm*** 1 and update N to form new training data $\{x_k^{i+1}, y_k^{i+1}\}, k = 1, 2, \cdots, N$.
8: **end if**
9: **end for**
10: $\{sx_k^i, sy_k^i\} \leftarrow \{x_k^i, y_k^i\}, k = 1, 2, \cdots, N$.

When using IPEM-online, we sparsify the training data set whether the APE_{AB}^{i} is less than 1% or not because IPEM has good fitting and sparseness ability. The main procedures of IPEM-online are shown in **Algorithm 6**.

Algorithm 6 IPEM-online.

Input:

The sparsified data set by using LSSVM-IPEM algorithm, $\{x_k, y_k\}, k = 1, 2, \cdots, N$;
The balises' number on the railway line, Num_{bal};

Output:

HST's location, the sparsified training data set: $\{sx_k^i, sy_k^i\}, k = 1, 2, \cdots, N$;

1: **for** i=1 to Num_{bal} **do**
2: Execute step 2 to step 3 of $\boldsymbol{Algorithm}$ 5.
3: **if** $APE_{AB}^{i} \geq 1\%$ **then**
4: Execute step 7 of $\boldsymbol{Algorithm}$ 4.
5: **end if**
6: Sparsify the training data $\{sx_k^i, sy_k^i\}, k = 1, 2, \cdots, N$, by using $\boldsymbol{Algorithm}$ 3 and update N to form new training data $\{x_k^{i+1}, y_k^{i+1}\}, k = 1, 2, \cdots, N$.
7: **end for**
8: $\{sx_k^i, sy_k^i\} \leftarrow \{x_k^i, y_k^i\}, k = 1, 2, \cdots, N.$

Algorithm 7 Hybrid-online.

Input:

The sparsified data set by using LSSVM-L_0NM algorithm, $\{x_k, y_k\}, k = 1, 2, \cdots, N$;
The balises' number on the railway line, Num_{bal};

Output:

HST's location, the sparsified training data set: $\{sx_k^i, sy_k^i\}, k = 1, 2, \cdots, N$;

1: **for** i=1 to 10 **do**
2: Execute step 2 to step 8 of $\boldsymbol{Algorithm}$ 5.
3: **end for**
4: **for** i=11 to Num_{bal} **do**
5: Execute step 2 to step 7 of $\boldsymbol{Algorithm}$ 6.
6: **end for**

For Hybrid-online, we combine L_0-norm-online with IPEM-online to balance the importance of the location accuracy and real-time response. In the initial ten adjacent balises, we use **Algorithm 5** to sparsify the training data set to obtain good sparsity and real-time performance. In the subsequent adjacent balises, we use **Algorithm 6** to sparsify the training data set to obtain good location accuracy performance. The main procedures of Hybrid-online are shown in **Algorithm 7**.

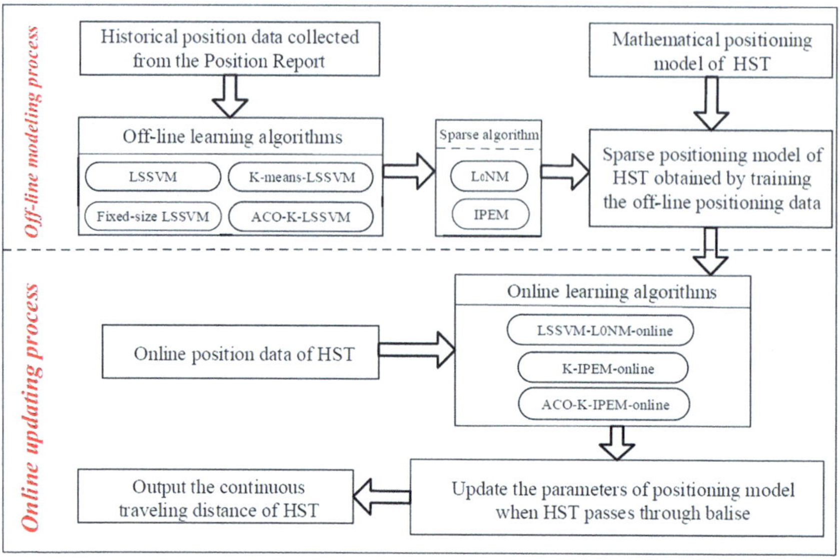

Fig. 4.8 Flowchart of intelligent localization method based on ACO and machine learning algorithms

4.2.3 Intelligent Train Localization Method Based on ACO and Machine Learning Algorithms

In order to improve the real-time performance on the premise of increasing the position accuracy of HST, the basic framework of intelligent positioning method based on intelligent algorithms is depicted in Fig. 4.8. In addition, the proposed method comprises two parts, such as off-line modeling process and online updating process.

For off-line modeling method, LS-SVM model is used to predict the continuous operational distance between two adjacent balises. In addition, k-means algorithm is used to better differentiate the train position data. When the clustering number of k-means algorithm increases, the over-fitting problem of single k-means is more serious. To solve this problem, ACO algorithm is integrated with k-means algorithm to adaptively optimize the clustering number of position data. In the operational process of HST, the wireless communication between HST and RBC will generate a large number of positioning data. If the LS-SVM model is directly trained by the original positioning data, the LS-SVM model will contain many support vectors, which will increase the calculation time for training and updating the LS-SVM model online. So, in order to apply the proposed method in online learning, sparse algorithm (IPEM) is applied to process classified positioning data after classifying the original positioning data set by k-means algorithm. For online updating process, to improve the adaptive ability of online learning model, the online learning algorithms are designed based on the IPEM and off-line positioning algorithms. To begin with, the training data set is classified by k-means or ACO-k-means algorithms. Then, the classified data will be sparsified by the proposed IPEM algorithm to improve the online real-time performance. Finally, for ensuring the real-time and the adaptive performance, the oldest positioning data is replaced by the latest positioning data to form new training data set.

A. Train Positioning Model Based on LS-SVM and K-Means Clustering Algorithms

Because position data contains different operational speed and traveling time interval, the LS-SVM prediction model should be established for different classes of position data. For distinguishing position data, k-means clustering algorithm, an unsupervised learning algorithm, is used to classify the position data and compute clustering centers after assigning clustering number for historical position data. Given a training data set $x^{(i)} \in \mathbb{R}^n$, $\forall i = 1, 2, \ldots, m$, cluster centroids are randomly initialized as $\mu_j \in \mathbb{R}^n$, $\forall j = 1, 2, \ldots, k$. m and k represent the number of training examples and the number of clusters we want to find, respectively. Then, cluster type $c^{(i)}$ of each training data is updated by Eq. (4.25) and cluster centroids μ_j are updated to the mean of the points assigned to it by Eq. (4.26).

$$c^{(i)} := \arg\min_{j} \left\| x^{(i)} - \mu_j \right\|^2, \quad \forall i = 1, 2, \ldots, m \tag{4.25}$$

$$\mu_j := \frac{\sum_{i=1}^{m} I\{c^{(i)} = j\} x^{(i)}}{\sum_{i=1}^{m} I\{c^{(i)} = j\}}, \quad \forall j = 1, 2, \ldots, k \tag{4.26}$$

The classification effect of k-means clustering algorithm can be measured by the following distortion function:

$$J(c, \boldsymbol{\mu}) = \sum_{i=1}^{m} \left\| x^{(i)} - \mu_{c^{(i)}} \right\|^2. \tag{4.27}$$

where $x^{(i)} \in \mathbb{R}^n$ represents a training example and m is the number of training examples. $\mu_{c^{(i)}}$ is the cluster centroid of class $c^{(i)}$ and $c^{(i)} = 1, 2, \ldots, k$. In order to avoid getting stuck in bad local minima, different random initial values for the cluster centroids μ_j are used to run k-means algorithms many times. Finally, out of all the different clustering found, the corresponding distortion $J(c, \boldsymbol{\mu})$ will be computed by Eq. (4.27) and the one that gives the lowest distortion $J(c, \boldsymbol{\mu})$ will be chosen. The detailed procedures of K-means-LS-SVM are shown in **Algorithm 8**.

Algorithm 8 Off-line position prediction model based on k mean clustering and LSSVM (K-means-LSSVM).

Input:
The existing data set, $\{x_i, y_i\}$, $i = 1, 2, \cdots, N$;
The balises' number on the railway line, Num_{bal};
The cluster number of training data set, k;
Output:
HST's position, the clustering of positioning data, absolute positioning error (APE_{AB});
Initialize cluster centroids $\mu_1, \mu_2, \cdots, \mu_k \in \mathbb{R}^n$ randomly.
repeat
for i=1 to m **do**
Compute the class type $c^{(i)}$ of each training data by Eq.(4.25).
end for
for j=1 to k **do**
Compute cluster centroids μ_j by Eq.(4.26).
end for
until convergence
Compute the distortion $J(c, \boldsymbol{\mu})$ of cluster centroid $\boldsymbol{\mu}$ by Eq.(4.27) when the iterative time is 50.
Choose the cluster centroid $(c^*, \boldsymbol{\mu}^*)$ gives the lowest distortion $J(c, \boldsymbol{\mu})$.
for j=1 to k **do**
for i=1 to m **do**
if $c^{(i)} == j$ **then**
Add the original position data sample $\{\boldsymbol{x}^{(i)}, y^{(i)}\}$ into the training data $\{\boldsymbol{cx}_j, sy_j\}$ of cluster centroid μ_j.
end if
end for
Given existing training data $\{cx_j^{(i)}, y_j^{(i)}\}$, $\mathrm{i} = 1, 2, \cdots, N_j$, linear systems (4.12) is solved to obtain optimal (γ_j, σ_j^2) and $(\boldsymbol{\alpha}_j, b_j)$ by considering RBF as the kernel function.
end for
for i=1 to n **do**
for j=1 to k **do**
$c^{(i)} \leftarrow \arg\min_j \left\|x^{(i)} - \mu_j\right\|^2$.
end for
end for
for i=1 to n **do**
if $c^{(i)} == j$ **then**
Compute the transient traveling distance by $y(\boldsymbol{x}) = \sum_{k=1}^{N} \alpha_j^* \boldsymbol{K}(\boldsymbol{x}, \boldsymbol{cx}_j) + b^*$ and APE_{AB} value.
end if
end for
return $\{\boldsymbol{cx}_j^{(i)}, y_j^{(i)}\}$ and APE_{AB}.

B. Train Positioning Model Based on LS-SVM and ACO-*K*-Means Algorithms

Traditional *k*-means algorithm can be used to cope with clustering problems with a known clustering number. However, the clustering number of position data is usually unknown for different train operational routes and conditions. In addition, if the clustering number is excessive, the fitting performance of single *k*-means algorithm will get worse for the testing data set. Therefore, the ACO algorithm [16] is introduced to solve the over-fitting problem of *k*-means by adaptively optimizing the number of clustering centers of position data. The detailed procedures are shown in **Algorithm 9**.

ACO algorithm is prone to stagnation phenomenon in the process of search. In other words, all solutions tend to be consistent when the search comes to a certain stage. For this reason, the solution space cannot be further searched and it is not conducive to find a better solution. So, the procedures are executed until the cluster centers are converged in Algorithm 9. The algorithm terminates when the variation sum for all cluster centers V during the updating process is less than the pre-specified value ε.

Since the ACO algorithm requires a long search time, the training data set is preprocessed by k-means algorithm before applying the ACO algorithm to speed up the convergence of the algorithm. The probability of incorporating class ω_i into class ω_j is determined by the amount of pheromone in path of class ω_i and class ω_j in the tth iteration process. In addition, the amount of pheromone in path of class ω_i and class ω_j is determined by the Euclidean distance d_{ij} of class ω_i and class ω_j and the cluster radius r. p_0 is the probability threshold value that incorporating class ω_i into class ω_j.

Algorithm 9 The position calculation model based on ACO, k-mean clustering and LSSVM (ACO-K-LSSVM).

1: Given sample data $\{X_i = (x_{i1}, x_{i2}, \cdots, x_{in})\}$, $i = 1,2 \cdots, N$, allocate M samples into N classes $(\omega_1, \omega_2, \cdots, \omega_N)$ by using k-means algorithm.

2: $V \leftarrow \infty$.

3: **while** $V > \varepsilon$ && $N! = 1$ **do**

4: **for** i=1 to $N-1$ **do**

5: **for** j=i+1 to N **do**

6: Compute the cluster center vectors of class ω_i that contains N_{ω_i} sample data by Eq.(4.28).

$$\overline{X^{(\omega_i)}} = \frac{1}{N_{\omega_i}} \sum_{k=1}^{N_{\omega_i}} X_k,\ (X_k \in \omega_i) \tag{4.28}$$

7: Compute Euclidean distance d_{ij} of class ω_i and class ω_j by the following equation:

$$d_{ij} = \left\| \overline{X^{(\omega_i)}} - \overline{X^{(\omega_j)}} \right\| = \sqrt{\sum_{k=1}^{n} \left(\overline{X_k^{(\omega_i)}} - \overline{X_k^{(\omega_j)}} \right)^2} \tag{4.29}$$

8: Calculate the amount of pheromone in path of class ω_i and class ω_j at time t.

$$\tau_{ij}(t) = \begin{cases} 1 & if\ d_{ij} \le r \\ 0 & if\ d_{ij} > r \end{cases} \tag{4.30}$$

where cluster radius $r = A + d_{min} + (d_{max} - d_{min}) \cdot B$, A and B are constant, d_{max} and d_{min} represent the maximum and minimum of the Euclidean distance d_{ij} calculated by step 3.

9: Compute the probability of incorporating class ω_i into class ω_j.

$$p_{ij}(t) = \frac{\tau_{ij}^{\alpha}(t)\eta_{ij}^{\beta}(t)}{\sum_{s \in S} \tau_{is}^{\alpha}(t)\eta_{is}^{\beta}(t)} \tag{4.31}$$

where $S = \{s | d_{sj} \le r, s = 1,2,\cdots,j-1,j+1,M\}$ represents a set of all classes whose distance is less than or equal to r, s represents a class number, M is the current total class number, $\eta_{is}(t)$ is weight parameter.

```
        if p_ij(t) ≥ p0 then
            Incorporate class ω_i into class ω_j.
            N ← N − 1.
            break.
        else
            break.
        end if
      end for
   end for
   Update the cluster center X^(ω_{i+1}) by Eq.(4.28).
   Compute the sum variation for all clustering centers, V ← Σ_{i=1}^N X^(ω_{i+1}) − X^(ω_i).
end while
Execute step 12 to step 29 of Algorithm 8.
return APE_AB.
```

C. Online Learning Algorithms for Train Positioning

Because the operational environment of HST is changing constantly, online learning algorithms are designed to improve the adaptive ability of the proposed intelligent positioning model. In order to update the positioning model during the operational process of HST, the latest positioning data set based on wireless communication will be added to the training data set when HST passes a balise. However, it is a time-consuming task to train the LS-SVM model when the positioning data is very large. In engineering practice, only a short traveling distance can produce a large amount of position report information. In addition, the latest added positioning data samples are not always better than the old ones. So, the training data set must be sparsified to improve the real-time performances of proposed intelligent positioning methods. Considering the introduced error of positioning data, the sample will be omitted from the original positioning data set when the corresponding sample bears the smallest error [15]. The introduced error of sample j described by $d(x_j)$ can be calculated by Eq. (4.21) when sample x_j is deleted from the original training data set. In Eq. (4.21), $\hat{y}_m(x_j)$ represents the evaluation of the output at iteration m of training sample x_j, α_k^m and b^m denote the multiplier for sample k and the bias at iteration m, respectively.

From Eq. (4.21), we can find that $d(x_j)$ is affected by more parameters besides α_j^m. IPEM is proposed in [11, 17] to obtain sparse LS-SVM model and the detailed procedures of IPEM can refer to [11, 17]. Based on the IPEM and off-line positioning algorithms, we design online learning algorithms in this section. The main procedures of online learning algorithms (*K-IPEM-online* and *ACOK-IPEM-online*) are shown in **Algorithm 10**. To begin with, the training data set is classified by k-means (**Algorithm 8**) or ACO-k-means (**Algorithm 9**). Then, the classified data will be sparsified by the proposed IPEM algorithm to improve the online real-time performance. Finally, for ensuring the real-time and the adaptive performance, the oldest positioning data is replaced by the latest positioning data to form new training data set. When the APE_{AB}^{i} is bigger than 1%, it is clear that the current LS-SVM model can't adapt to the change of environment of HST. So, the positioning model will be updated by the new formed training data.

Algorithm 10 Online positioning algorithm (*K-IPEM-online* and *ACOK-IPEM-online*).ψ

Input:

The existing data set, $\{x_i, y_i\}$, $i = 1, 2, \cdots, N$;

The balises' number on the railway line, Num_{bal};

Output:

HST's position;

Sparsified training data set, $\{sx_k^i, sy_k^i\}$, $i = 1, 2, \cdots, N$;

Absolute positioning error, APE_{AB}.

1: Given the existing data set, $\{x_i, y_i\}$, $i = 1, 2, \cdots, N$, obtain the classified positioning data $\left\{cx_j^{(i)}, cy_j^{(i)}\right\}$, $i = 1, 2, \cdots, N_j$, $j = 1, 2, \cdots, k$, by **Algorithm 8** or **Algorithm 9**.

2: Sparsify the classified positioning data $\left\{cx_j^{(i)}, cy_j^{(i)}\right\}$ by using IPEM sparse algorithm and update N_j to form sparsified data $\left\{sx_j^{(i)}, sy_j^{(i)}\right\}$, $i = 1, 2, \cdots, N_j$, $j = 1, 2, \cdots, k$.

3: Given the sparsified data $\left\{sx_j^{(i)}, sy_j^{(i)}\right\}$, $i = 1, 2, \cdots, N_j$, $j = 1, 2, \cdots, k$, solve linear system (4.12) to obtain optimal parameters $(\gamma_j^i, \sigma_j^{2i})$ and $(\alpha_j^i, \sigma_j^{2i})$ by 10-fold cross-validation for j class type.

4: Compute the cluster center for each class $\overline{X^{(\omega_j)}}$, $j = 1, 2, \cdots, k$ by Eq.(4.28).

5: **for** i=1 to Num_{bal} **do**

6: Compute Euclidean distance d_{ij} of input data $tx^{(i)}$ and class centroid $\overline{X^{(\omega_j)}}$ by Eq.(4.29).

7: Predict the HST's position by using the LSSVM model established in the step 3 and compute APE_{AB}^i value with (α_j^i, b_j^i) for j class type.

8: **if** $APE_{AB}^i \geq 1\%$ **then**

9: Add the latest position report data between ith adjacent balises in $\left\{sx_j^{(i)}, sy_j^{(i)}\right\}$, $i = 1, 2, \cdots, N_j$, $j = 1, 2, \cdots, k$, and update N_j to form new training data $\left\{sx_j^{(i)}, sy_j^{(i)}\right\}$, $i = 1, 2, \cdots, N_j$, $j = 1, 2, \cdots, k$.

10: **end if**

11: Update the cluster center $\overline{X^{(\omega_j)}}$, $j = 1, 2, \cdots, k$, by Eq.(4.28).

12: **end for**

13: **return** APE_{AB}.

4.2.4 Speed Measurement and Positioning of Multi-sensor Information Fusion

Multiple types of sensors are used for multi-sensor information fusion to obtain high positioning accuracy. Currently, the commonly used onboard speed measurement and positioning sensors mainly include the following types [1]:

(1) Speed sensor: speed sensor positioning method is the most commonly used train positioning method. At present, the research of train positioning method based on speed sensor focuses on the detection of position information during train idling and sliding.
(2) Acceleration sensor: in order to overcome the imprecise positioning of the speed sensor caused by train idling and sliding, the train usually uses the acceleration information of the train acceleration sensor to modify the speed measurement and positioning information of the train.
(3) Doppler radar: it is a positioning and speed measuring device using doppler effect. The radar was developed and put into use in the 1960s. Since the 1970s, with the development of large-scale integrated circuits and digital processing technology, pulsed doppler radar has been widely used in airborne early warning, navigation, missile guidance, satellite tracking, aircraft reconnaissance, and so on. This kind of radar is small in size, light in weight, and high in reliability, so it is also suitable for train positioning and speed measurement (mostly used for urban rail).
(4) Gyroscope: gyroscope is an instrument that can accurately determine the direction of moving objects. It is an inertial navigation instrument widely used in modern aerospace, navigation, aerospace, and defense industries. The gyroscope is compact in structure, high in sensitivity, and reliable in operation. It is very suitable to be used as an auxiliary positioning tool for train positioning.

Due to the lack of robustness of a single sensor system, the occasional failure may lead to the failure of the whole system, or even disastrous consequences for the train positioning system. In order to make the system have fault detection and fault isolation ability, experimental system uses a hierarchical structure to combine different sensors and construct the χ^2 type of residual statistics. χ^2 distribution is used to detect the fault of sensor. In this structure, all fusion nodes provide navigation solutions, and the higher the level of nodes, the higher the precision. When a sensor of higher level node fails, the navigation solution of the lower level node is output. It is found that the positioning accuracy of the query balise positioning method is not only related to the layout interval of the query balise, but also related to the location of the query balise. Since the train is prone to idling and sliding in the curve section, if the query balise is installed after the curve section along the running direction of the train, the positioning error will be better improved.

In addition, after decades of development, multi-sensor information fusion technology has achieved certain results in all aspects. At present, most researches on multi-sensor information fusion are carried out for specific problems in specific application areas, that is, based on the types of problems, intuitive fusion criteria are established to form the optimal fusion scheme. According to the JDL function model, Hall and Iinas present three fusion system structures, such as centralized fusion, autonomous fusion, and hybrid fusion. Dasarathy gives the structure of three fusion systems according to the input and output characteristics: flexible system structure, self-improvement system structure and finite decision input and fusion structure. Foster believes that the structure of the fusion system is centralized, decentralized, and mixed. The centralized structure takes data from all sensor platforms and processes the data in the fusion center. The decentralized structure requires each sensor platform to process local information and make situation estimation.

However, because there are few query balises arranged along the Qinghai–Tibet railway, the positioning and speed measurement errors of multi-sensor information fusion are difficult to be eliminated. In addition, the environment of Qinghai–Tibet railway is complex and the temperature difference is large, which also increases the measurement errors of onboard sensors.

4.2.5 Satellite Navigation and Positioning

Train positioning methods based on satellite navigation include GPS positioning method, navigation prediction method (**DR**), and GPS/DR integrated positioning method:

(1) GPS Positioning System: Global positioning system (GPS) is a satellite-based wireless navigation system shown in Fig. 4.9.
In order to eliminate the influence of SA policy, ionosphere and troposphere on positioning accuracy, differential global positioning system (DGPS) has attracted much attention from many scholars in recent years. DGPS technology is to place a GPS receiver at a reference station with known geographical location, calculate the position data of satellite launch, and then send the correction information to other receivers in the region. In this way, its positioning accuracy can be greatly improved. High positioning accuracy is extremely important for high-speed trains or high-density subways. GPS system can provide Standard Positioning Service (SPS) and Precise Positioning Service (PP). SPS service is used to track, schedule, and control trains.
In the train positioning system, high precision, low cost of GPS devices can be chosen as the user part of the GPS receiver. As long as the GPS signal is available, it can provide the absolute coordinates of the train, especially by using high-precision GPS satellite signal carrier phase for positioning. For the GPS satellite positioning has good low-frequency response characteristics, the precision is still very high in thousands of meters of distance, and the error does not accumulate over time. In addition, GPS can also be used to collect geographic location data and draw high-precision electronic map of high-speed railway lines. Moreover, the combination of GPS technology and mobile com-

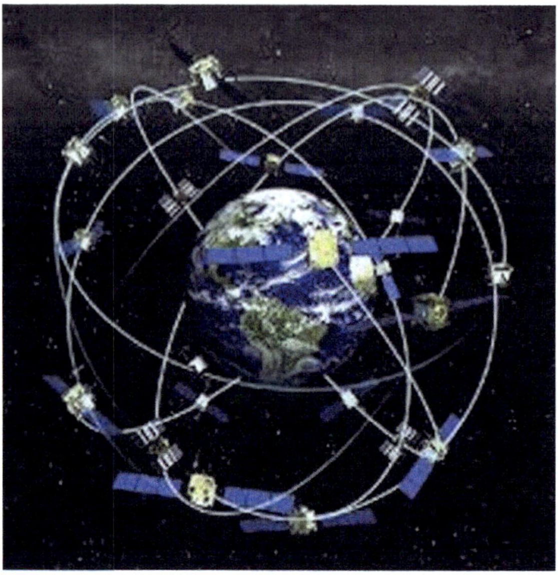

Fig. 4.9 Global positioning system

munication system can realize train positioning tracking. The GPS receiver on the train collects data and sends it to the dispatching center through the mobile communication system. Then, the electronic screen of the dispatching center can display the running state of the train in real time.

However, GPS positioning will be affected by obstacles such as "urban canyons" formed by tunnels, mountains, forests, and station buildings. GPS signal will be weakened or even disappeared, which will make the positioning tracking ineffective. At this point, we need to adopt the dead reckoning method.

(2) Dead reckoning (DR) positioning method: DR positioning method is based on the correction of relative position, and its current position is calculated by using the coordinates of known starting point and the running speed and direction of the train. The system mainly includes displacement sensors and course sensors. Since the inertial sensors completely rely on the onboard equipment to complete the navigation tasks, they do not have any photoelectric contact with the outside world and are not limited by climatic conditions, so they are widely used in the transportation field.

(3) GPS/DR integrated positioning method: the navigation position estimation system composed of inertial sensors has better high-frequency and low-frequency characteristics, while GPS is opposite to the performance of it. If a combined positioning system is constructed using the advantages of both, the precision, performance, and reliability of the whole system can be improved greatly.

The total length of Qinghai–Tibet railway is 1956 km. Due to the high altitude, it is not comfortable to live and work for a long time. So, only seven of the 45 stations are manned. In addition, because the Qinghai-Tibet railway has a large departure interval and low train density, train control does not require very accurate positioning information. For the reasons given above, China railway sector used the global positioning system for the first time in the locomotive monitoring and management of Qinghai–Tibet railway. At the same time, the digital dispatch control and management are applied in the opening operation of Qinghai–Tibet railway by integrating GPS with the wireless digital mobile communication technology.

Due to special environmental, geographical, and cultural factors, the Qinghai–Tibet railway has three requirements for train operation control: to realize the minimum travel time of all passenger trains, the "maintenance-free" equipment, and the basic "impersonal" management along the route. In order to meet the above requirements, based on GPS satellite positioning, the Qinghai–Tibet railway apply ITCS, which set automatic block, station interlocking control, the overspeed protection control for train operation at a suit, and use the virtual block, such as GSM-R wireless communication way in the device to improve traffic density under the premise of ensure train safety. By using ITCS, the maintenance workload can be minimized and high extensibility and compatibility can be obtained to meet the harsh environmental conditions along the Qinghai–Tibet railway under the condition of reducing trackside equipment.

4.3 GPS Positioning Requirements

Because the Qinghai–Tibet railway is mainly built in the empty plateau area, there are few large buildings, and the proportion for crossing high mountain tunnel is relatively low, which is favorable for GPS real-time positioning. Thus, GPS is mainly used to quickly obtain three-dimensional coordinates of spatial objects in real time.

GPS data is collected in real time by GPS receiver of the information receiving and processing system of onboard positioning system. On the one hand, according to the operational condition of online train, the dispatcher manually or the system automatically carries out the call communication on the relevant train. The information receiving and processing system of onboard positioning system sends the train's current position, running speed, operational direction, train number, and other relevant information of train to the dispatching center. On the other hand, when the train needs to make an emergency call to the dispatching center, it can communicate with the dispatching center in real time and upload the train's position, speed and running direction and other relevant information, which can be used for the command and decision-making of the dispatching center.

For train positioning, satellite positioning receivers are installed at the head and tail of the train, respectively. Normally, the relative positions of the two receivers are fixed. When the train decoupling or other circumstances causes a change in such relative position, the dispatcher and driver can be reminded to confirm, so as to assist train integrity inspection.

Table 4.2 The accuracy standard of GPS measurement

	Fixed error a (mm)	Proportional error coefficient b
B	≤ 8	≤ 1
C	≤ 10	≤ 5
D	≤ 10	≤ 10
E	≤ 10	≤ 20

4.3.1 Accuracy Requirement

Since GPS satellite positioning is a positioning method under relative motion, the accuracy of positioning will be limited. However, as long as GPS signal is available, the absolute position information of the train can be provided by GSP positioning system. Even at a long distance, the carrier phase is used for positioning, and thus the accuracy will be high because the accumulation of time will not affect the positioning error. Due to the special environment of Qinghai–Tibet railway, the accuracy of positioning has an important influence on the operation safety of train and vehicle scheduling.

According to the accuracy, GPS railway measurement should be divided into four grades: B, C, D, and E. The precision of chord length of the adjacent points of GPS network at all levels is expressed by Eq. (4.32) and should comply with the provisions of Table 4.2.

$$\sigma = \sqrt{a^2 + (b * d)^2} \tag{4.32}$$

where σ is the standard deviation (mm), a fixed error (mm), b proportional error coefficient (mm/km), d distance between adjacent points (km).

To improve the accuracy of GPS network measurement, the fixed error a should meet the requirements of Table 4.2. For class b, the proportional error coefficient b shall meet the requirements of twice as listed in Table 4.2, and class C to E shall still meet the requirements of Table 4.2.

In accordance with the principle of optimal design, the GPS network should be deployed by considering its purpose, the expected accuracy, the visibility of satellites during operation, the reliability of the results, the number of receivers and traffic and other logistics conditions involved in the operation.

A GPS network should consist of one or more independent observation rings. When the reliability and precision requirements of the network are high, triangular or geodetic quadrangular networks should be adopted. When the accuracy requirement is low, quadrangular network, traverse network, alignment line, or mixed network including these networks can be adopted. It is generally not advisable to use a single base point.

The connection between GPS network synchronous graphs should be edge-linked or network-linked. When the accuracy requirement is not high, it can also use the point-linked network, but the whole network positioning results should be enhanced to check for preventing coarse difference.

The GPS network should be connected with a nearby national ground point or an existing GPS point for joint measurement. The total number of joint measurement points should not be less than three and, in special cases, not less than two. When the number of joint measurement points is two, it should be distributed at both ends of the network. When the number of joint measurement points is three or more, it is advisable to distribute evenly in the network.

When using GPS for altitude control measurement, GPS network points should be properly connected to the elevation points according to the accuracy requirements and the terrain conditions of the measurement area, in addition to the national level point. The **elevation** of the joint elevation point can be measured by grade level measurement (including five level measurement) or other methods equivalent to its precision.

4.3.2 Real-Time Requirement

The train position information is provided to the dispatching center in real time, so as to realize the real-time monitoring of the whole train and ensure the safety of the train. The train position information continuously provided by GPS covers the whole line (station and operational zone) of the train operation, which can be used to assist the adjustment of operation plan and accurately monitor and control the train in real time.

4.3.3 Economic Requirement

The economic requirement should consider not only the cost of equipment investment, but also the maintenance cost of equipment. The equipment investment cost mainly includes the investment of GPS positioning equipment, namely the GPS receiver and differential error information receiver installed at both ends of the train. Since there is no need to set ground equipment for GPS satellite positioning method, a large amount of installation and maintenance costs are concentrated on GPS receiver and differential error information receiver. Therefore, on the premise of accurate real-time positioning, how to adopt low-precision GPS receiver and information receiver to reduce the maintenance cost of equipment will be the focus of research.

4.3.4 Interference Immunity Requirement

GPS satellite positioning is easy to be interfered by obstacles, which reduces its positioning accuracy. The bad weather of Qinghai–Tibet railway has great influence on its performance. In addition, this method is also subject to the performance of satellite and space station equipment. Therefore, it is also important to improve the anti-interference capability of GPS satellite positioning.

4.4 Key Scientific Problems

The improvement of GPS positioning performance comes from two aspects: One is the improvement of satellite system and the reception technology. The other is the improvement of data processing methods and refinement of various models. In the GPS positioning system of Qinghai–Tibet railway, the key problem is how to deal with the large amount of the collected GPS data. GPS signals are affected by various error sources from transmitting to receiving. The improvement of GPS data processing method is of great practical significance to make full use of and excavate the potential function of the original GPS observation data, which can improve the positioning accuracy of train control system in Qinghai–Tibet railway.

According to the regulations of the China Railway Corporation (former ministry of railways), GPS observation data should meet the following requirements:

(1) The observation results should meet the basic technical requirements of dispatching command and corresponding GPS surveying operations.
(2) Items of measuring manual record should be complete and effective.
(3) The test of processing result of synchronous edge and the test result of subnet composed by the synchronous edge. The closure error of synchronous ring and the independent ring shall conform to the provisions of this regulation.
(4) The quality of the baseline vector shall meet the following requirements:

- The results of any two periods on the same side should be no more than 2 times of the nominal precision of the receiver.
- When the baseline quality factor meets the software requirements, the baseline processed by batch software is not subject to synchronous loop test. Otherwise, the closure error test of the synchronous ring composed of baselines that operated manually should be conducted, and the closure error shall conform to the provisions of Eq. (4.33).

$$\begin{cases} w_x \leq \frac{\sqrt{n}}{5}\sigma \\ w_y \leq \frac{\sqrt{n}}{5}\sigma \\ w_z \leq \frac{\sqrt{n}}{5}\sigma \\ w \leq \frac{\sqrt{n}}{5}\sigma \end{cases} \tag{4.33}$$

where $\{w_x, w_y, w_z\}$ is the closure error of coordinate component, w closure error of the full length of the ring, n the number of edges of the closed loop, σ accuracy specified for the corresponding grade (calculated by the mean length of the ring).

- A closed loop consisting of several independent edges or synchronous edges solved by different mathematical models shall meet the requirements of Eq. (4.34):

$$\begin{cases} w_x \leq 3\sqrt{n}\sigma \\ w_y \leq 3\sqrt{n}\sigma \\ w_z \leq 3\sqrt{n}\sigma \\ w \leq 3\sqrt{3n}\sigma \end{cases} \tag{4.34}$$

When the length difference between the long edge and short edge is large in the closed loop, the accuracy (σ) for each edge should be calculated in accordance with regulations of the accuracy of both the length and the level. In addition, the precision of closure error of the **ring** (loop) should be computed according to the law of error propagation for substituting $\sqrt{n}\sigma$ in Eqs. (4.33) and (4.34).

- The correction difference of baseline observed repeatedly shall conform to the provisions of Eq. (4.35):

$$\mathrm{ds} \leq 2\sqrt{2}\sigma \tag{4.35}$$

where ds is the correction difference of baseline observed repeatedly, σ is the precision specified for the corresponding grade (calculated by the length of the edge).

- Independent ring test shall be conducted for all baseline solutions. In general, if the baseline doesn't participate in the closed loop difference test, the baseline should not be contained in the network.

Under the above standards, GPS data processing mainly includes GPS data reduction, error data detection, and multi-GPS track information fusion.

4.4.1 GPS Data Reduction

After obtaining a large number of GPS data, an important problem is to use effective reduction algorithm to express the track simply and efficiently, so as to reduce the storage space and improve the matching efficiency of electronic map, under the condition of controlling the matching error within the allowed range. Railway track contains straight track and curve track, the representation of straight track is relatively simple, while the representation of curved track on electronic map is a difficult problem. Currently, the commonly used methods are NURBS representation, Bezier curve representation, etc. However, the representation of such curves will lead to the increase of data storage, especially the complexity of the corresponding map matching algorithm [18].

The actual curved track is in an asymptotic shape with a large radius of curvature. It has been found by previous studies that the error between the piecewise line and curve track can be controlled within a certain range as long as fewer points are taken. There are two kinds of track error represented by broken line: horizontal error and vertical error. The horizontal error is the maximum orthogonal projection distance of the broken line from the track. The vertical error is the difference between the length of the track and the length of the broken line. So, the points on the track are taken in order, and the adjacent broken line is used to approximate the curve track. In this book, the combined mathematical model of data reduction is derived, and then a heuristic algorithm is proposed [19]. The performance of the proposed algorithm is analyzed and compared with the field GPS data.

4.4.2 Error Data Detection

By using a lot of GPS data with high precision, to generate the electronic map of rail transit automatically is a development direction of GIS for rail transit. Whether the electronic map of rail transit is correct or not depends on the accuracy, integrity, and confidentiality of the measured GPS data. The accuracy of data refers to the precision of GPS data itself, which is guaranteed by differential GPS technology. The integrity of data and the confidentiality of records are mainly dependent on the patience and meticulous care of field surveyors.

However, some error data will inevitably appear in a large number of measured GPS data. The most basic method to detect the error data is to manually observe the scatter diagram of the measured data step by step. However, this manual method has a heavy workload and is prone to errors due to fatigue. Therefore, it is very important to develop an automatic detection algorithm to detect and judge the location of GPS error data and then manually determine it, which can greatly reduce the workload and ensure the correctness of judgment.

4.4.3 Multiple GPS Trajectory Information Fusion

For train control system, precise GPS electronic map plays an important role in train safety operation, precise positioning, train scheduling, and other aspects. High-precision track electronic map is a kind of navigation information source which is not affected by the outside world and has high stability. It can not only make the train know all kinds of information about the route, but also effectively correct the train positioning error and improve the positioning accuracy and reliability. GPS electronic map needs to obtain accurate GPS data measured by GPS receiver. The accuracy of

GPS receiver is mainly affected by orbit error, satellite clock difference, ionospheric error, tropospheric error, carrier phase cycle jump and multi-path error, etc. Moreover, the train system itself has a complex electromagnetic environment, which makes the high-precision GPS receiver used in the train expensive. Therefore, by using low-cost GPS receiver, a great deal of low-precision trajectory can be obtained through multiple measurements. Then, the automatic generation of high-precision trajectory can be realized through optimization algorithm, which can greatly reduce the cost [6].

Since it is difficult to give the mathematical equation of railway track, the fitting error can be better described by using orthogonal distance as measurement than vertical distance, which is similar to principal component analysis. Hastie extended the principal component analysis algorithm to non-linearity in 1988 and proposed the principal curve theory and algorithm. On this basis, some scholars proposed the main curve algorithm composed of segmented polylines and automatically recognized the digital images, which achieved good results. There are many fixed Points of Interest (PI) along the railway, such as the position of signal machine and turnout switch machine. PI points are some fixed constraint, and high position precision of PI is required for the safe running of the train. However, traditional main curve algorithms can't deal with the fixed constraint points' data in railway track. Therefore, to solve this problem, this book integrates multiple low-precision GPS tracks to achieve high-precision positioning on the basis of the previous methods. The proposed algorithms are verified with both the simulated data under complex shapes and the measured railway data.

4.5 Summary

This chapter first describes the important role of the positioning system in CTCS-3 level train control system, as well as the composition of the positioning system and the principle of GPS satellite positioning. The second section summarizes three common positioning methods in the train control system: query balise positioning, speed measurement and positioning based on multi-sensor information fusion, satellite navigation positioning, and analyzes and compares the advantages and disadvantages of the three positioning methods. The third section mainly analyzes the requirements of GPS in train positioning, namely the requirements of accuracy, real-time performance, economy, and anti-interference. The fourth section analyzes the key scientific problems of GPS in train positioning, mainly including hardware configuration, research and development and data processing.

References

1. Ning B, Tang T, Li KC, Dong HR. High-speed train control system. Beijing: Science Press; 2012.
2. Dong H, Ning B, Cai B, et al. Automatic train control system development and simulation for high-speed railways. IEEE Circuits Syst Mag. 2010;10(2):6–18.
3. Zhu AH, Li B, Yang L. Research on positioning technology and combined positioning system for high-speed train. Chin Railw. 2013;5:59–63.
4. Chen DW, Gao Q, Pei LJ. Study on digital track map based on graph theory and oriented object. J China Railw Soc. 2012;34(6):64–8.
5. Liu J, Cai BG, Tang T, et al. Fault-tolerant algorithm design of low-cost integrated train positioning system. J China Railw Soc. 2011;33(1):39–46.
6. Guo BQ, Tang T, Zhou DT. Map aided positioning method used in train locating application. J China Railw Soc. 2007;29(6):44–7.
7. Liu J, Cai BG, Tang T, et al. Train integrated positioning system based on GPS and inertial measurement unit. China Railw Sci. 2010;31(1):123–9.
8. Kang SW, Liu XJ. Research on train location method based on multi-sensor information fusion. Railw Comput Appl. 2014;23(1):14–7.
9. China's Ministry of Railways. Railway survey regulations of global positioning system (TB10054-97). Beijing: China Railway Press; 1997.
10. Tang JG. Analysis for train control system of Qinghai-Tibet railway based on GPS and GSM-R. J Transp Syst Eng Inf Technol. 2010;10(2):190–4.
11. Cheng R, Song Y, Chen D, Cheng L. Intelligent localization of high-speed train using LSSVM and online sparse optimization approach. IEEE Trans Intell Transp Syst. 2017;18:2071–84.
12. Chen D, Han X, Cheng R, Yang L. Position calculation models by neural computing and online learning methods for high-speed train. Neural Comput Appl. 2016;27:1617–28.
13. Chen D, Wang L, Li L. Position computation models for high-speed train based on support vector machine approach. Appl Soft Comput. 2015;30:758–66.
14. Suykens JAK, De Brabanter J, Lukas L, et al. Weighted least squares support vector machines: robustness and sparse approximation. Neurocomput. 2002;48(1–4):85–105.
15. de Kruif BJ, de Vries TJA. Pruning error minimization in least squares support vector machines. IEEE Trans Neural Netw. 2003;14(3):696–702.
16. Dorigo M, Maniezzo V, Colorni A. Ant system: optimization by a colony of cooperating agents. IEEE Trans Syst Man Cybern Part B (Cybern). 1996;26:29–41.
17. Cheng R, Song Y, Chen D, Ma X. Intelligent positioning approach for high speed trains based on ant colony optimization and machine learning algorithms. IEEE Trans Intel Transp Syst. 2018. To be published.
18. Jia XZ, Chen DW. Study on information fusion algorithm for multiple GPS railway track data. J China Railw Soc. 2011;33(9):72–5.
19. Chen DW, Cai BG, Wang J, Tang T. Mathematical model and algorithm of data reduction for railway GPS data. J China Railw Soc. 2008;30(4):116–9.

Chapter 5
Error Data Detection

5.1 Overview

There are a large number of measurement data points of GPS that are distributed on Qinghai–Tibet railway with a length of 1142 km. Due to various measurement reasons, there are inevitably some measurement errors. It is very important to develop a method to detect possible errors from all data points, and thus the reliability of GPS data can be improved by modifying or re-measuring them. Four error modes based on expert knowledge exist in the measurement data, including redundant measurement, sparse measurement, back-and-forth measurement, and large angle change. For the four error modes of the measurement data, four algorithms need to be developed to detect the possible errors of the corresponding data points. In order to remove repeated error data points and effectively display possible errors with different algorithms, an integrated error detection method is proposed by reasonably assembling four algorithms. After evaluating the error detection method with four performance indexes, the method can be verified with six GPS track data sets between seven railway stations of QTR. In this way, it is possible that thirty-eight sections of data of some consecutive points are wrong, and 14 of them have been confirmed by the measurement experts. This method has a detection rate of 100% and lasts less than half an hour, far less than the manual workload of 94 h. The verification results show that the detection method is effective, not only reducing the workload, but also ensuring its accuracy by integrating domain expert knowledge to make the final correct decision.

Satellite positioning has many advantages, such as low cost, good real-time performance, and no cumulative error [1]. It is often used in automobile navigation system [2] or electronic map [3]. At present, satellite positioning has been used to track train location, besides radio frequency system [4] and track circuit. The use of satellite positioning system can reduce the installation and maintenance cost of trackside equipment [5]. So, the European Union has launched many projects using satellite positioning for low-density railways, such as GADEROS [6] and RUNE [7]. In the USA, the railway in Michigan adopts the GPS positioning incremental train

D. Chen and R. Cheng, *Intelligent Processing Algorithms and Applications for GPS Positioning Data of Qinghai-Tibet Railway*, https://doi.org/10.1007/978-3-662-58970-0_5

control system [8]. The results of these projects indicate that satellite positioning has a better cost-performance ratio for low-density railways. In China, the first GPS-guided train control system was used on the QBR in 2006, and no track circuits were used. On the world's highest railway, the cost of wayside equipment and maintenance is greatly reduced [9].

High-precision digital track map (DTM) is the basis of accurate train positioning. With the help of DTM, positioning error is reduced and positioning reliability is improved [10]. Obviously, the GPS track data generated by DTM can be easily used for train positioning. In order to generate highly reliable DTM, high-precision GPS track data points need to ensure the integrity and confidentiality of data records. On the Qinghai–Tibet railway, different GPS technology is used to realize the high accuracy of GPS track data points. The integrity of the record, however, depends largely on rigorous, patient, and careful measuring workers. With the length of 1142 km, a large number of GPS satellite positioning data points on Qinghai–Tibet railway are needed. But, some measurement errors are inevitable due to various reasons for measurement.

Railway track contains linear track and curvilinear track. Describing railway track by curve will increase storage space, especially the complexity of algorithm [11]. However, the railway track can be composed of several sections of curves. As long as the error is controlled within a certain range, the railway track can be obtained according to the endpoints of the curve section [12].

In engineering practice, the basic method applied by engineers for finding measurement errors is to check the data points in sections, which obviously wastes time, and is easy to lose some errors. Therefore, in order to automatically find possible errors and improve the reliability of data, an effective error detection method should be developed. In different fields, expert knowledge plays a key role in the analysis of degraded terrain, seismic design of reinforced concrete buildings, detection of fraudulent communication networks, etc. Therefore, some measurement rules summarized by experts are reasonable and can be used to design error detection algorithms, which can improve work efficiency and ensure the correctness of the judgment.

In this chapter, some error detection algorithms based on expert knowledge will be developed and thus the errors in GPS track data points in some railway stations of QTR can be discovered. In Sect. 5.1, in addition to a brief description of the QTR, six GPS data sets between seven railway stations are also briefly described. In Sect. 5.2, four error data patterns and four corresponding error detection algorithms are summarized through expert knowledge. In Sect. 5.3, an integrated error detection method is proposed, which can not only display the error effectively, but also remove the possible repeated errors detected by different algorithms. Then, the performance of the method can be evaluated through four performance indicators. In Sect. 5.4, the threshold is set by the expert knowledge and the error detection method can be used to find some possible errors in the GPS track data set. The calculation results of six data sets are presented and compared with the manual judgment method. Some conclusions and applications are presented in Sect. 5.5.

The Qinghai–Tibet railway, a high-altitude railway linking Golmud, Qinghai province, to Lhasa, Tibet, was officially opened by President Hu Jintao on July

Fig. 5.1 A train on the QTR after inauguration

1, 2006. Figure 5.1 shows the official operation diagram of QTR. Figure 5.2 shows the major railway stations and mains mountain ranges along the QTR.

The Qinghai–Tibet railway includes Tanggula railway station at an altitude of 5072 m and Fenghuoshan tunnel at 4905 m. Tanggula railway station is the highest railway station in the world, and the Fenghuoshan tunnel is the highest railway tunnel in the world. For the 1142 km distance of QTR, about 500,000 GPS data points with precision of 1 cm can be obtained through the differential global positioning system technology. The distance between each two adjacent points is between 1.5 and 3 m, and the average distance is about 2.5 m.

This chapter studies six data sets between seven railway stations. Due to the GPS trajectory data of the QTR have strict security requirements, we need a special software to transform the latitude, longitude, and altitude of GPS data into cartesian coordinate system x–y–z and set the data set of first railway station as the origin of coordinates.

The change in height (z coordinate) is much smaller than that in the x–y plane when the railroad gradient is very small. By preprocessing the survey data set, we found that the distance between two adjacent points was between 1.5 and 3 m. However, when all data points are projected onto the z-axis, the average distance between two adjacent points in the z coordinate is between 0.0011 and 0.023 m. Therefore, compared with the change of x–y plane, the change of z-axis can be ignored.

In order to clearly show the data points, this chapter only shows x–y plane of GPS measurement data. These data sets are shown in Figs. 5.3, 5.4, and 5.5, respectively. The length of each part is between 20 and 30 km. The data sets of six sections

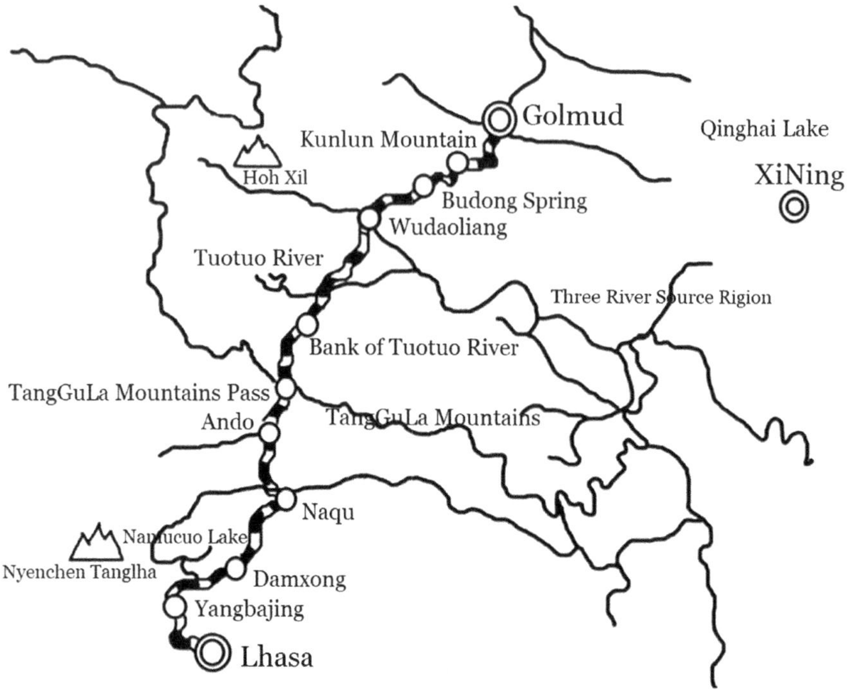

Fig. 5.2 Main railway stations and main mountains along the QTR

are defined as data set $\text{dataset}_i = [\text{station}_i, \text{station}_{i+1}], \quad (i = 1, 2, \ldots, 6)$, and respectively, contain 9279, 9768, 10,257, 11,233, 9475, and 8595 data points. From the beginning to the end, the measurement rules are used to record these data points and the DGPS technology is used to make these GPS data points reach 1 cm accuracy.

5.2 Four Error Modes and Error Detection Algorithms

5.2.1 Four Error Patterns Based on Expert Knowledge

By consulting some GPS measurement experts and field measurement engineers, we found that four error patterns often occur during the measurement and recording process. From the experience of actual measurement, although the measured density is smaller in the linear track than that in the curved track, the distance between two adjacent data points is still in a certain range. Therefore, the four error patterns are described below.

- As shown in Fig. 5.6, error mode 1 (point p_i) consists of three cases, namely redundant measurement, density measurement, and repetitive recording.

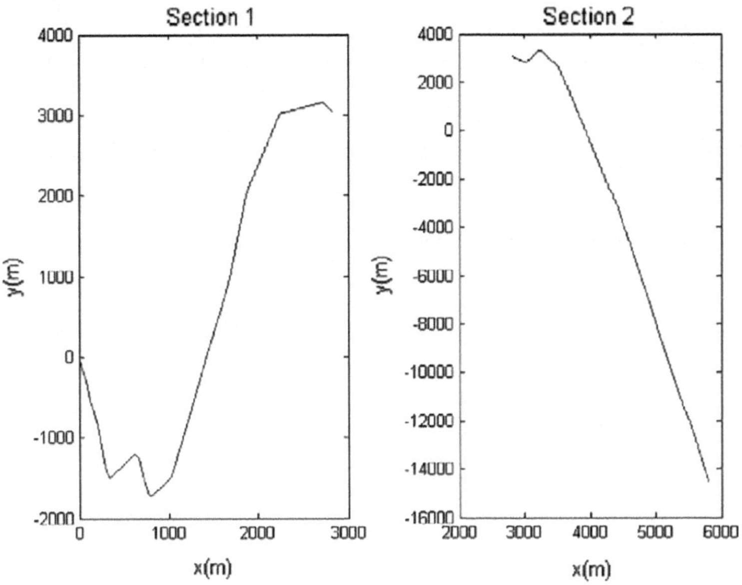

Fig. 5.3 GPS data points in *x*–*y* plane for Sects. 5.1 and 5.2

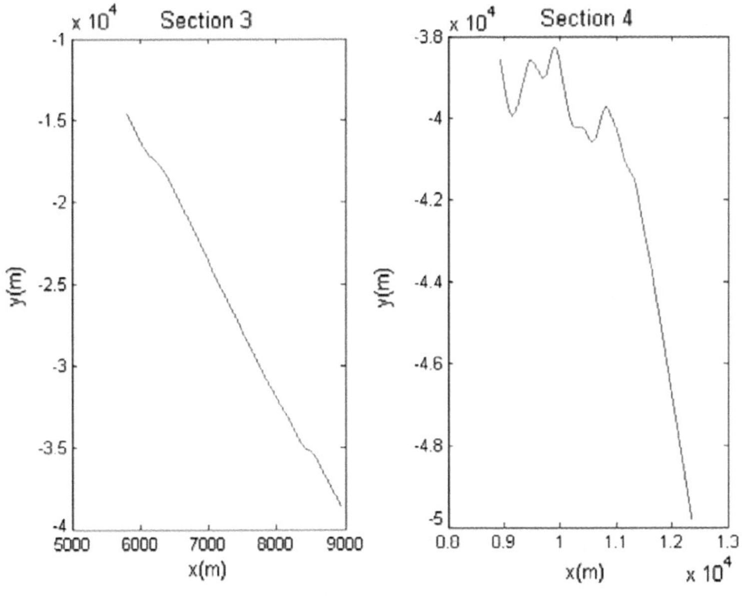

Fig. 5.4 GPS data points in *x*–*y* plane for Sects. 5.3 and 5.4

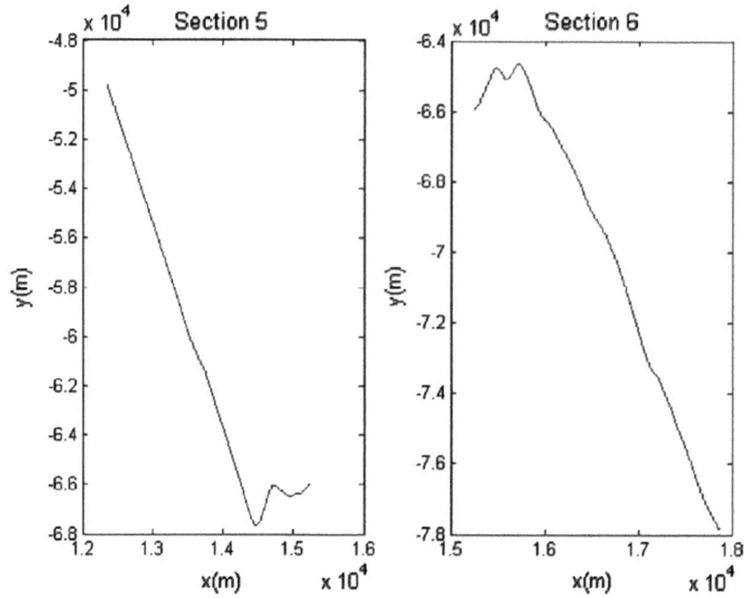

Fig. 5.5 GPS data points in x–y plane for Sect. 5.5

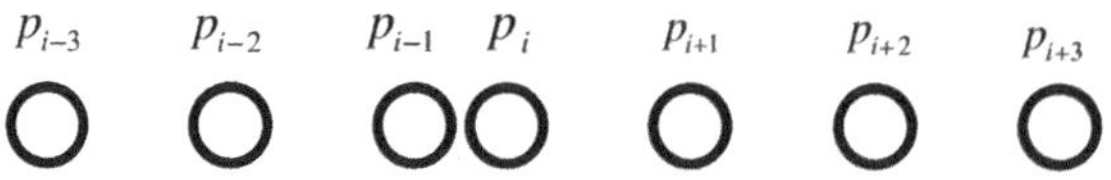

Fig. 5.6 Error pattern 1

Fig. 5.7 Error pattern 2

- Error mode 2 is known as sparse measurement, missing points, and forgotten records. As shown in Fig. 5.7, the points between p_{i-1} and p_i belong to error mode 2.
- Error mode 3 is defined as a round measurement or circular measurement. As shown in Fig. 5.8, p_i and p_{i+1} belong to error mode 3. In the measurement, there is a back and forth measurement or missing point due to false error. Since the latter segment is before the first, this type of error can lead to a lot of redundancy points in very short railway segments and make DTM unreasonable.
- Error mode 4 refers to the large angle change caused by the significant deviation of data points from railway track. As shown in Fig. 5.9, the point p_i belongs to this type. As a railway track, whether it is a straight line or a curve with a large

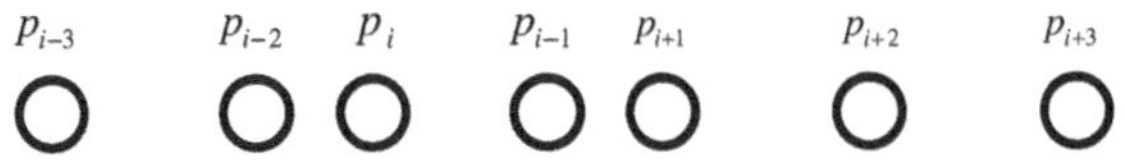

Fig. 5.8 Error pattern 3

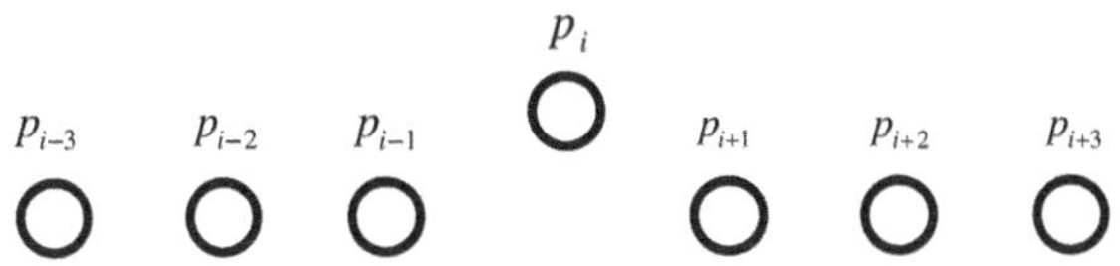

Fig. 5.9 Error pattern 4

radius of curvature, the angle change between two adjacent lines is very small. If a data point is clearly far from the track, the change in angle will be large.

It is important to note that the first two error patterns are not very serious errors. Error mode 1 only has redundant data points that can be removed to save storage space. Error mode 2 has no adverse effect on the straight rail, but the measuring engineer can be reminded to add some measurement data points if necessary.

5.2.2 Four Error Detection Algorithms

To make the error detection algorithms accurate and easy to understand, we define some symbols as follows:

- p_i is ith point; x_i, y_i and z_i are the coordinates of p_i, so $p_i = (x_i, y_i, z_i)$;
- N is the number of all points in the data set;
- dist_i is the distance between p_i and p_{i+1}; dist2_i is the distance between p_i and p_{i+2};
- l_i is the linear connection between p_i and p_{i+1}; θ_i is the slope between l_i and l_{i+1};
- $\langle p_i, p_j \rangle$ is the inner product of p_i and p_j.

For these four error modes, four algorithms are proposed to detect them. According to the error mode 1, dist_i can be used to find possible errors. As shown in Eq. (5.1), if dist_i is smaller than threshold dist_{min}, it is a false or redundant data points.

$$\text{dist}_i = \sqrt{(x_{i+1} - x_i)^2 + (y_{i+1} - y_i)^2 + (z_{i+1} - z_i)^2} < \text{dist}_{\text{min}} \quad (i = 1, 2, \ldots, N - 1) \tag{5.1}$$

According to error mode 2, errors can also be found through dist_i. As shown in Eq. (5.2), if dist_i is greater than the threshold dist_{max}, it is a false data or missing data points.

$$\text{dist}_i = \sqrt{(x_{i+1} - x_i)^2 + (y_{i+1} - y_i)^2 + (z_{i+1} - z_i)^2} > \text{dist}_{\text{max}} \quad (i = 1, 2, \ldots, N - 1) \tag{5.2}$$

For error mode 3, the errors can be found through dist2_i, as shown by Eq. (5.3). Under normal circumstances, dist_i should be multiplied by 2. If the back-and-forth measurements occur, dist2_i will be small or even less than dist_{min}. The threshold value $\text{dist2}_{\text{min}}$ is approximately equal to dist_{min} times 2.

$$\text{dist2}_i = \sqrt{(x_{i+2} - x_i)^2 + (y_{i+2} - y_i)^2 + (z_{i+2} - z_i)^2} < \text{dist2}_{\text{min}} \quad (i = 1, 2, \ldots, N-1) \tag{5.3}$$

For error mode 4, errors can be found by using the changing angle θ_i between two adjacent lines.

$$\theta_i = \frac{180}{\pi} * \cos^{-1} \frac{\langle p_{i+1} - p_i, p_{i+2} - p_{i+1} \rangle}{\text{dist}_i \times \text{dist}_{i+1}}, \quad (i = 1, 2, \ldots, N-1) \tag{5.4}$$

where $cos^{-1}x$ is the inverse cosine of x in radians. Under normal conditions, the θ_i is very small in a straight rail track. Therefore, the absolute value of θ_i must be less than the threshold θ_{max}. Otherwise, an incorrect data point may appear in Eq. (5.5).

$$\text{abs}(\theta_i) > \theta_{\text{max}} \quad (i = 1, 2, \ldots, N-1) \tag{5.5}$$

The threshold values can be determined by experienced experts, such as dist_{min} in formula (5.1), dist_{max} in Eq. (5.2), $\text{dist2}_{\text{min}}$ in Eq. (5.3) and θ_{max} in Eq. (5.5). Therefore, expert knowledge has great influence on error detection algorithm. In addition, the scatter diagram of dist_i, dist2_i and θ_i can be given by experts by examining threshold.

5.3 Integrated Error Detection Methods and Performance Index

5.3.1 Integrated Error Detection Algorithm Based on Four Error Detection Algorithms

If a possible error is found at the p_i point by any algorithm, for the convenience of observation, we collect five points near p_i as a set of suspicious points [p_{i-2}, p_{i-1}, p_i, p_{i+1}, p_{i+2}] (if such nearby points exist). Because some errors can occur in different error modes, for example, error mode 3 will result in error mode 4. In order to reduce the repeat count of possible errors in GPS data points, it is necessary to combine the four algorithms to find all possible errors and delete repeated points. After combining error points, possible errors are divided into segments and each segment contains multiple consecutive data points.

Through four stages of extension, combination, deletion, and display, four algorithms are integrated into an integrated error detection method, which can effectively detect and show the errors hidden by GPS data points on the Qinghai–Tibet Railway.

Then, the possible error points detected by the integrated error detection method will be accepted or rejected by the measurement expert system based on the expert's knowledge and experience. Therefore, the integrated expert knowledge error detection method not only improves the work efficiency, but also guarantees the correctness of the results. It is important to note that the error detection method only finds possible errors in each segment, including some that are confirmed as true errors by experts.

5.3.2 *Performance Index*

In order to effectively evaluate the effectiveness of error detection methods, we proposed the following four performance indicators:

1. Detection Rate (DR)

$$\mathrm{DR} = \frac{S_r}{S_h} \tag{5.6}$$

where S_r is the number of truly incorrect sections discovered by the detection method, and S_h is the number of truly incorrect sections discovered by GPS measurement experts. If DR is much closer to 1, the effectiveness of this method will be better tested.

2. False Alarm Rate (FAR)

$$\mathrm{FAR} = \frac{S_f}{S_a} \tag{5.7}$$

where S_f is the number of false errors, and S_a is the number of possible errors. Since possible errors include real errors and false errors, we can get Eqs. (5.5–5.8). The closer the FAR is to 0, the easier it will be for human judgment. Therefore, FAR reflects the accuracy of the method. The lower the FAR, the higher the accuracy of the method.

$$S_a = S_f + S_r \tag{5.8}$$

3. Efficiency of Measurement (EM)

$$\mathrm{EM} = \frac{T_h}{T_m} \tag{5.9}$$

where T_h is the running time of manual judgment, T_m is the running time of error detection method. The larger the EM, the higher the measurement efficiency of the method.

4. Reliability of Measurement (RM)

$$\mathrm{RM} = 1 - \frac{N_r}{N} \tag{5.10}$$

where N_r is the number of real errors detected by error detection, N is the number of all points in the data set. The closer RM approaches to 1, the better the quality of measurement.

5.4 Calculation Results and Analysis

5.4.1 Set the Threshold

After consulting with railway measurement expert, the four thresholds can be set to the following values:

(1) $\mathrm{dist}_{\mathrm{min}} = 1.5\,\mathrm{m}$, $\mathrm{dist}_{\mathrm{max}} = 3\,\mathrm{m}$;
2) $\mathrm{dist2}_{\mathrm{min}} = 3\,\mathrm{m}$;
3) $\theta_{\mathrm{max}} = 1^{\circ}$.

For determining the threshold, another method is to observe the dist_i, $\mathrm{dist2}_i$, and θ_i by scatter diagram. As shown in Figs. 5.10, 5.11 and 5.12, three scatter diagrams were obtained from $\mathrm{dataset}_1$ to verify the accuracy of expert experience. The scatter diagram of dist_i shown in Fig. 5.10 conforms to the expert knowledge ($\mathrm{dist}_{\mathrm{min}} = 1.5\,\mathrm{m}$, $\mathrm{dist}_{\mathrm{max}} = 3\,\mathrm{m}$). The scatter diagrams of $\mathrm{dist2}_i$ and θ_i are also conform to the above expert knowledge ($\mathrm{dist2}_{\mathrm{min}} = 3\,\mathrm{m}$), ($\theta_{\mathrm{max}} = 1^{\circ}$), respectively. More possible errors are detected when we set $\mathrm{dist}_{\mathrm{max}}$ and θ_{max} to be small or $\mathrm{dist}_{\mathrm{min}}$ and $\mathrm{dist2}_{\mathrm{min}}$ to large. DR and FAR will increase, which means high efficiency will lead to low accuracy. Otherwise, fewer possible errors will be detected and the FAR and the DR will decrease, which means that the high accuracy will lead to a decrease in the effectiveness of the detection. In practice, it is more important to find all real mistakes than to reduce false ones. Therefore, DR is set as 100% and high FAR is acceptable.

5.4.2 Calculation Results and Performance Analysis

After all possible errors in the six data sets are found by the four error detection algorithms, the error detection method needs to detect a total of 38 segments possible errors by extending, combining, deleting, and displaying. Fourteen of them are guaranteed by GPS measurement experts. The processing time of each data set by using error detection algorithm is about five minutes, and the total time is about half

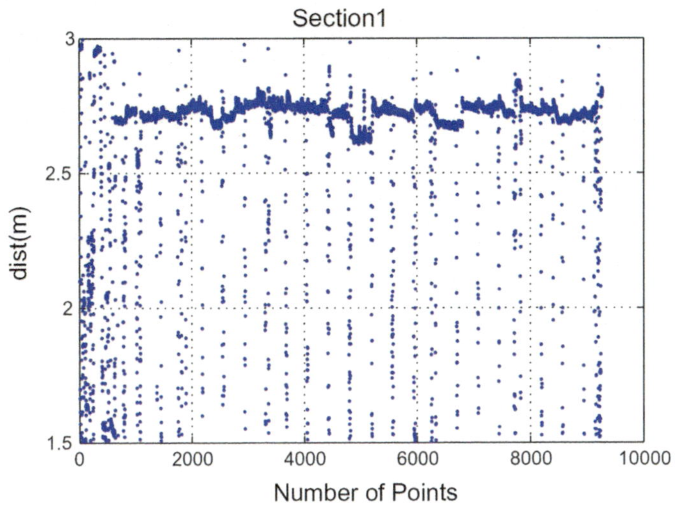

Fig. 5.10 The scatter plot of $dist_i$

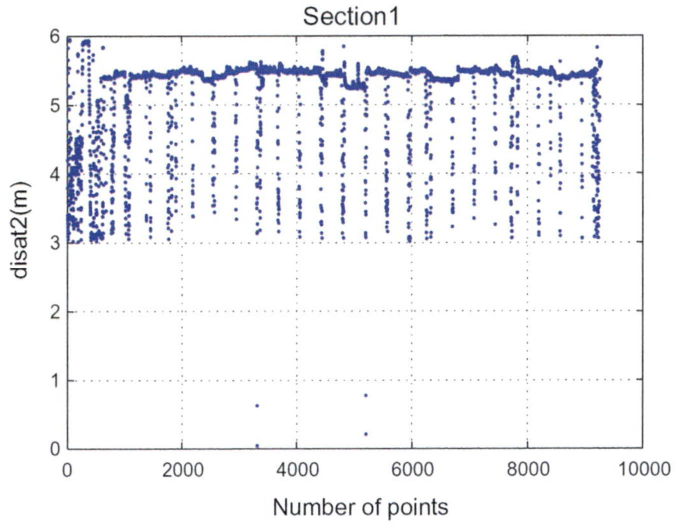

Fig. 5.11 The scatter plot of $dist2_i$

an hour. In the same way, it takes about 94 h for a measurement engineer to find errors one by one, and the 14 segments that they find errors are contained in the 38 segments that were discovered by the error detection method.

Table 5.1 shows the detailed calculation results for each data set. From Table 5.1, we can find that the integrated error detection method proposed in this chapter is very effective, because DR for all data set is equal to 100%, and the average value of EM is 207. In addition, as the average of RM is 99.86%, the reliability quality of

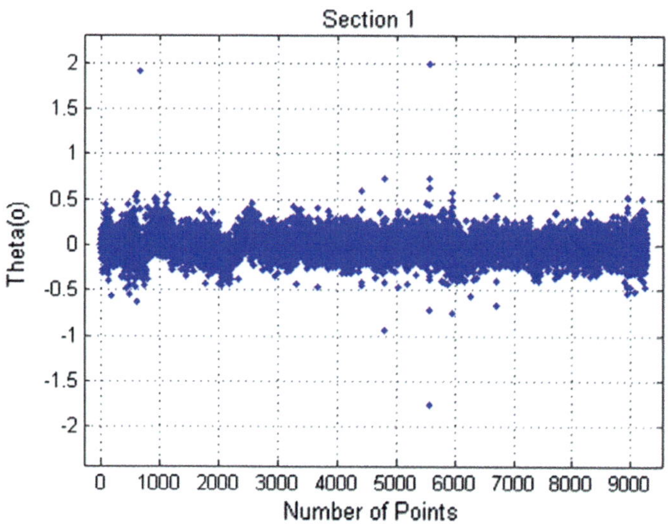

Fig. 5.12 The scatter plot of θ_i

Table 5.1 The calculated results of the six data sets by using the integrated error detection method

Section	N	S_a	S_h	DR (%)	FAR (%)	T_m(min)	T_h(min)	EM	N_r	RM (%)
1	9279	4	2	100	50	2.8	900	321	13	99.88
2	9268	5	1	100	80	3.5	960	274	6	99.95
3	10,257	1	1	100	0	2.2	980	445	6	99.95
4	11,233	14	6	100	57.1	8.8	1020	116	35	99.69
5	9475	10	3	100	70	7.4	930	126	17	99.85
6	8595	4	1	100	75	2.6	860	331	7	99.94
Total	58,607	38	14	100	63.2	27.3	5650	207	84	99.86

the test is very high. To ensure that all possible errors are found, the average value of FAR is 63.2%.

In the 14 segments that are really wrong, two segments contain error mode 1, one segment contains error mode 2, seven segments contain error mode 3, and four segments contain error mode 4. Various error modes were selected and shown in Figs. 5.13, 5.14, 5.15 and 5.16, indicating that many errors were hidden in a large number of GPS data points during the measurement. It should be noted that error mode 1 and error mode 2 have less errors than error mode 3 and error mode 4.

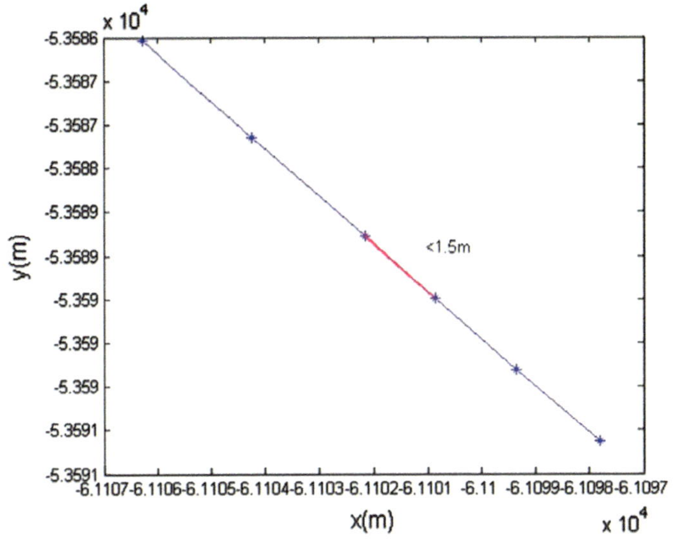

Fig. 5.13 Error pattern 1 found by the integration method

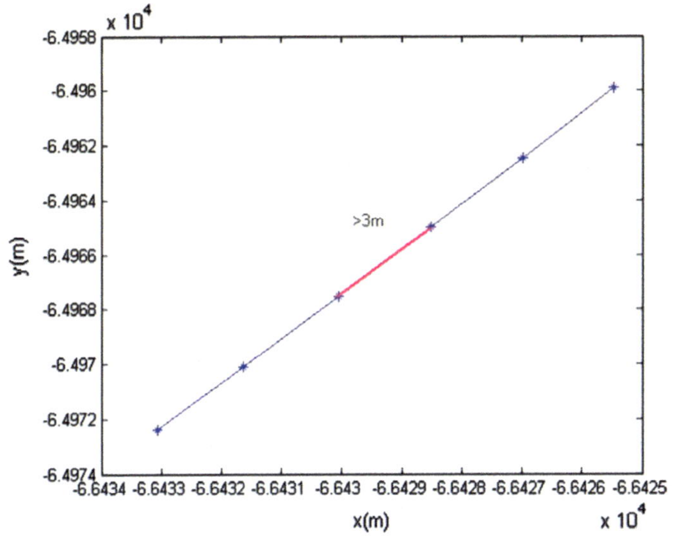

Fig. 5.14 Error pattern 2 found by the integration method

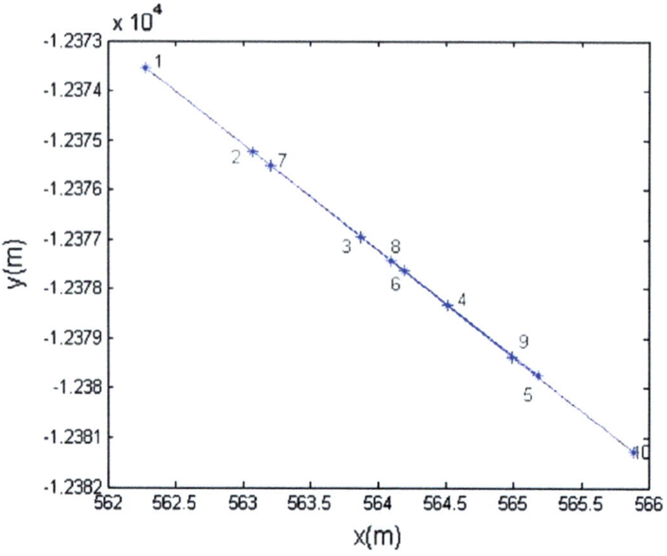

Fig. 5.15 Error pattern 3 found by the integration method

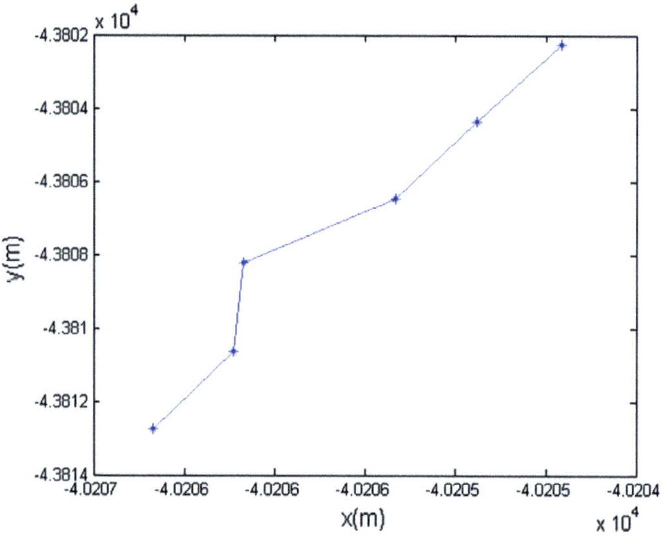

Fig. 5.16 Error pattern 4 found by the integration method

5.5 Brief Summary

For the Qinghai–Tibet railway, some measurement errors inevitably exist in a large number of GPS measurement data points. To reduce manual workload, improve work efficiency, and ensure the reliability of GPS track data points, effective error detection methods are necessary to find possible errors automatically and quickly. In this chapter, four error modes are summarized from the knowledge of measurement experts and four algorithms are designed to detect them, respectively. In addition, some repeated errors are removed by combining the four algorithms and the possible errors are effectively displayed.

The computation results show that the error detection method is effective for detecting the fault GPS data points since it can find almost all errors. Especially, it saves a lot of time and greatly improves detection efficiency. While FAR is a bit higher, it helps prevent the absence of real errors. For the proposed error detection approach, it is worth noting that expert knowledge in the field plays an important role not only in determining the threshold, but also in detecting the true error after sectionally displaying the possible errors. Therefore, this method is a comprehensive method that combines expert knowledge and the computing power of computer. After detecting errors, experts increase the reliability of GPS track data by removing redundant points or giving suggestions, which need to be measured again and will provide great support for follow-up work, such as generating digital track maps or designing train positioning algorithms.

According to theoretical research, a detection software is developed and thus those hidden possible errors can be automatically detected on the Qinghai–Tibet railway. From the practical application of field testing, we find that the software can quickly and effectively find out the errors hidden in a large amount of GPS data, which increases the reliability of GPS data on Qinghai–Tibet railway. Therefore, the integrated error detection method proposed in this paper can guarantee the following work, such as the generation of digital track map of high-precision train positioning system, which can be of great help to the operation of Qinghai–Tibet railway.

References

1. Blomenhofer H. GNSS in the 21st century-the user perspective. Acta Astronaut. 2004;54:965–8.
2. Skog I, Handel P. In-car positioning and navigation technologies—A survey. IEEE Trans Intell Transp Syst. 2009;10(1):4–21.
3. Zhang J, Chen D, Kruger U. Adaptive constraint K-segment principal curves for intelligent transportation systems. IEEE Trans Intell Transp Syst. 2008;9(4):666–77.
4. Santos AJD, Soares AR, de Almeida Redondo FM, et al. Tracking trains via radio frequency systems. IEEE Trans Intell Transp Syst. 2005;6(2):244–58.
5. Raymond G, Marais J, Berbineau M. Innovation brings satellite-based train control within reach. Railway Gaz Int. 2004;160(12):835–7.
6. Urech A, Pérez Diestro J, Gonzalez O. GADEROS, a GALILEO demonstrator for railway operation system. Data Syst Aerosp. 2002;509:442–7.

7. Albanese A, Marradi L, Labbiento G, et al. The RUNE project: The integrity performances of GNSS-based railway user navigation equipment. In: ASME/IEEE 2005 Joint Rail Conference. American Society of Mechanical Engineers; 2005. p. 211–218.
8. Baker J, Clennan J. ITCS: A new approach to increasing capacity. Int Railway J. 2005;45(10):46–7.
9. Li K. The project of ITCS signal control system of Qinghai-Tibet Railway. Chin Railways. 2005;7:31–6.
10. Simsky A, Wilms F. GNSS-based failsafe train positioning system for low-density traffic lines based on one-dimensional positioning algorithm. Conserv Matters. 2004 (March).
11. Gu YH, Tjahjadi T. Coarse-to-fine planar object identification using invariant curve features and B-spline modeling. Pattern Recogn. 2000;33(9):1411–22.
12. Gao G, Cai B. Research on the automatic electronic map generation algorithm for the train supervision system. J China Railway Soc. 2006;1:63–70.

Chapter 6
Data Reduction

6.1 Overview

Because satellite positioning has many advantages, such as low cost, good real-time performance and no accumulative error [1], it is widely used in automobile navigation [2] and electronic map software [3]. In addition, satellite positioning technology is also applied to train positioning and tracking by replacing the track circuit and radio frequency positioning technology that have been used in the past [4]. The adoption of satellite positioning technology can greatly reduce the installation and maintenance costs of the wayside equipment [5]. In recent years, the European Union has launched many low-density railway construction projects using satellite positioning technology, such as GADEROS [6] and RUNE [7]. In addition, an enhanced train control system using satellite positioning technology has been put into operation in Michigan railway in the USA for a long time [8]. The above projects show that the satellite positioning technology has a good cost-performance ratio on the low-density railway. In China, GPS positioning technology was first applied in the train control system of QTR line to realize zero existence of track circuit on the world's highest railway and greatly save the cost of operation and maintenance [9]. Relying on high-precision digital track map (DTM), the positioning accuracy and reliability of the train can be improved [10]. Because DTM is generated by a large amount of GPS track data, the application of effective data compression algorithm is very important in the huge satellite positioning data. Within the acceptable error range, excellent data compression algorithm can be selected for the purpose of drawing DTM with as few data points as possible.

Railway track can be divided into straight track and curved track. The curve representation [11] can be used to represent curved track, which will lead to the increase of data capacity and the complexity of corresponding map matching algorithm. Due to the large curvature radius of the curved rail and the smooth connection between the straight rail and the curved rail, the curved rail can be approximated by connecting some key points on the curved rail when the approximation error is controlled

D. Chen and R. Cheng, *Intelligent Processing Algorithms and Applications for GPS Positioning Data of Qinghai-Tibet Railway*, https://doi.org/10.1007/978-3-662-58970-0_6

within an acceptable range [12]. Therefore, some links between the key points on the railway track are often selected to approximate the actual track line.

6.2 Reduction Model of GPS Data

The idea of GPS data reduction in railway applications is described above. In other words, some key data points are selected from the collected GPS data set to approximate the actual track line, so as to achieve better results with as few points as possible. The selected data point constraint is that the orthogonal distance from each data point to the corresponding line is less than the following error constraint. Mathematical model of GPS data reduction [13] is conducted as follows:

$$\min \sum_{i=1}^{N} t_i \tag{6.1}$$

$$J = \{i_1, \ldots, i_m\} \subseteq \{1, \ldots, N\} \tag{6.2}$$

$$1 = i_1 < i_2 < \ldots < i_{m-1} < i_m = N \tag{6.3}$$

$$\operatorname{dist}\left(L\left(i_j, i_{j+1}\right), k\right) \leq \delta, \quad \text{if}\, i_j < k < i_{j+1} \tag{6.4}$$

$$\begin{cases} t_i = 1, & \text{if}\, i \in J \\ t_i = 0, & \text{if}\, i \notin J \end{cases} \tag{6.5}$$

$$\operatorname{dist}(L(i, j), k) = \sqrt{\|p_k - p_i\|_2^2 - \frac{\langle p_k - p_i, p_j - p_i \rangle^2}{\|p_j - p_i\|_2^2}} \tag{6.6}$$

Equation (6.1) is the objective function; N represents the total number of data sets to be simplified; J represents the set of selected key points; δ represents the error constraint of the ith data point; $L(i, j)$ represents the line between p_i and p_j; $\operatorname{dist}(L(i, j), k)$ represents the orthogonal distance from p_k to $L(i, j)$. Except for the first and last two points, the remaining $N-2$ points can be included as the key points to be selected in the J set. In these $N-2$ points, if p_i is selected, then t_i is equal to 1, otherwise t_i is equal to 0.

There are 2^{N-2} possibilities in the set to be selected, which makes it difficult to obtain the optimal solution of such big data problem by enumerating. Taking only a section of the Qinghai–Tibet railway as an example, the value of N has reached several thousand. Since Eq. (6.6) is non-linear, data reduction is a large-scale non-linear data combination problem. In order to make the algorithm highly interpretable, we give two relevant definitions as follows:

$\overline{p_i p_j}$ represents the line from p_i to p_j, for $i_j < k < i_{j+1}$, $\operatorname{dist}\left(L\left(i_j, i_{j+1}\right), k\right) \leq \delta$ means that the data points meet the error constraints.

$\overline{p_i p_j}$ has a deviation γ, if $\text{dist}(L(i_j, i_{j+1}), k) = \gamma$ when $\max\{i\} < k < j$, the deviation can be represented as $D(i, j) = \gamma$.

6.3 Algorithm and Performance Index

6.3.1 Algorithm 1: The Reduction Algorithm Based on Prediction Concept

The basic idea of the reduction algorithm based on prediction concept is to search from one end of the railway to the other end. After selecting the first point (which means $i_1 = 1$), then the second point i_2 is chosen, the algorithm terminates until the last point is selected. All chosen data points are selected from the set $\{2, 3, \ldots, N\}$. In this process, key points are selected as many as possible. The algorithm terminates when the selected number reaches N, or until the error after i_2 is greater than δ. The procedures of the algorithm are described as follows:

Step 1: given N points, set $\delta > 0$.
Step 2: $k = 1, i_k = 1$.
Step 3:

```
while k < N do
    if D(i_k, N) ≤ δ, then k = k + 1, add i_k into the selected set J;
end
```

$J = \{i_1, \ldots, i_m\}$ is the solution obtained from Algorithm 1. In the calculation process of Algorithm 1, it is not difficult to find that the calculation number of this algorithm is at least two times. The maximum calculation number is defined as $s = \max_j\{i_{j+1} - i_j\}$ for $j \in \{1, 2, \ldots, m-1\}$. The number of calculation steps required by Algorithm 1 is represented by the following equation:

$$o(sN) \approx o\left(\frac{N^2}{m}\right) < o(N^2) \tag{6.7}$$

6.3.2 Algorithm 2: The Reduction Algorithm Based on Dichotomy

Algorithm 1 only focuses on local data, while Algorithm 2 is a heuristic algorithm based on dichotomy. The advantage lies in taking the overall characteristics of data sets into account. The basic idea of Algorithm 2 is whether the data set needs to be divided into two smaller sets. It is necessary to divide the data set into two subsets if $D(i, N) > \delta$ items are included. For each subset, perform the same steps until all subsets do not need to be split. This is like the generation of a binary tree, the entire data set can be taken as a root node and a subset as a leaf node. The calculation process of Algorithm 2 is as follows:

Step 1: given N points, set $\delta > 0$.
Step 2: $J = \{1, N\}, m = 2$.
Step 3: the elements in J are set in this order: $1 = i_1 < i_2 < \ldots < i_{m-1} < i_{\mathrm{m}} = N$.
Step 4:

```
for i = 1: size(J)-1
    initialEle = J(i); endEle = J(i + 1);
    for j= initialEle: endEle
        if D(i_j, i_{j+1}) ≤ δ, break;
        elseif D(i_j, i_{j+1}) > δ, then K = j, add the K into J, and return to Step 3;
    end
end
```

$J = \{i_1, \ldots, i_m\}$ is the solution obtained from Algorithm 2. The maximum calculation number is defined as $s = \min_j \{ i_{j+1} - i_j \}$ for $j \in \{1, 2, \ldots, m-1\}$. The number of calculation steps required by Algorithm 2 is represented by the following equation:

$$o\big(N * \log_2(N/s)\big) \approx o\big(N * \log_2 q\big) \tag{6.8}$$

Since $q = N/s$ is much less than N, the computational complexity of Algorithm 2 is almost linear.

6.3.3 *Algorithm 3: The Reduction Algorithm Based on Breadth-First Strategy*

The above two algorithms do not optimize the data reduction problem well, so we propose a shortest path method to solve the data reduction problem. If $p_i p_j$ satisfies the bounded error constraint δ, we use a directed arc to connect from point p_j to point p_i. For this particular problem, we set the weight of each arc to 1.

When all lines are selected, a weighted directed graph is formed. Matrix A is the judgment matrix of the weighted directed graph. If $\overline{p_i p_j}$ meets the error constraint condition of δ, then $A(i, j) = 1$. Obviously, looking for the shortest path from point p_1 to point p_N is equivalent to looking for the minimum number of points that satisfy the formula (6.2)–(6.6). Therefore, the data reduction problem is transformed into the problem of finding the shortest path in the directed graph with the same weight 1. The worst case is that every line segment generated in a directed graph needs to be checked. For point p_i, we should check the line segment $\overline{p_i p_j}$ using the projection calculation method of $k(k = j - i - 1)$ in formula (6.6). Therefore, for all line segments $p_i p_j$ $(j = i + 2 : N)$, we need to carry out the $(N - i - 1)(N - 1)/2$ steps' inspection. So, the total number of calculations for all line segments $\overline{p_i p_j}(i = 1 : N - 2, j = i + 2 : N)$ is $\sum_{i=1}^{N-2}(N - i - 1)(N - i)/2$, which is the same thing as $\frac{N^3}{6} - \frac{N^2}{2} + \frac{N}{3}$. Because it is going to take a lot of computing time to detect all the line segments when N is greater than or equal to 5000. As a result, we put forward a kind of 2δ technology to reduce the computing time and save the storage space.

If the line segment $p_i p_j$ does not satisfy the error constraint δ, the part exceeding the deviation constraint will be detected. We check $p_i p_{j+k}$ sequentially until $D(i, j + k)$ is greater than 2δ. If $D(i, j + k)$ is greater than 2δ, let us go ahead and check $p_{i+1} p_{i+3}$. In this way, besides p_j, another point can be found to satisfy the error constraint condition of δ. Since the rails are mostly straight rails or curved rails with large radius of curvature, it is almost impossible to find a point of p_{m+k}, so that $D(i, m + k)$, $k \geq 1$ and $D(i, m)$ are both greater than 2δ. Many classic algorithms, such as Dijkstra and Flord, cannot solve different types of shortest path problems [14]. However, the advantage of breadth-first algorithm lies in the same weight of the directed arc in the weighted digraph. The process of Algorithm 3 based on breadth priority algorithm is as follows:

Step 1: $k = 0, i = 0, V(0) = \{p_i\}$.
Step 2: If $i = N$, go to Step 6. Otherwise, move to Step 3.
Step 3: if $p_i \notin V(k)$, then $i = i + 1$ and return to the Step 2. Otherwise, $j = i + 1$ jump to Step 4.
Step 4: if $p_i \notin V(k) \cup V(k - 1) \cup \ldots \cup V(0) \&\& D(i, j) \leq \delta$, then $V(k + 1) = V(k + 1) \cup p_i$, take p_i as the front point of p_j, and jump to Step 7. Otherwise, let $j = j + 1$ and jump to Step 5.
Step 5: if $j = N$, then $i = i + 1$ and return to the Step 2. Otherwise, $j = i + 1$ jump to Step 4.
Step 6: let $k = k + 1$, $V(k + 1) = \emptyset$, $i = 1$, and return to Step 3.
Step 7: if $j = N$, then the algorithm terminates. Otherwise, let $j = j + 1$ and return to Step 4 to continue the calculation.

Finally, the shortest path obtained from the previous point is found by reversely starting from p_N to p_1. As 2δ method, the V method applied in Step 4 also reduces the projection calculation. Each element in set $V(k)$ has the same shortest path from the initial point p_1.

The distance between two adjacent railways is about 5 m. Theoretically, if the distance is less than 2.5 m, the requirement of distinguishing adjacent railways can be achieved. In this chapter, to ensure security, the value of δ is set as 0.5 or 1 m. The following five evaluation indexes are used to evaluate the performance for different algorithms:

(1) The efficiency of the data reduction: $r(\%)$

$$r = 100 * \frac{N - m}{N} \tag{6.9}$$

where m represents the number of the chosen critical data points, N is the number of all data points.

(2) The missing degree of the data length: $e_l(\text{‱})$

$$e_l = 1000 * \left(1 - \frac{\sum_{j=1}^{m-1} \left\| p_{i_{j+1}} - p_{i_j} \right\|_2}{\sum_{i=1}^{N-1} \| p_{i+1} - p_i \|_2} \right) \tag{6.10}$$

(3) e_a (m) is defined as the average of all orthogonal distances from the data point to the corresponding line segment.

(4) e_m (m) is defined as the maximum of all orthogonal distances from the data point to the corresponding line. It is used to check whether the formula (6.4) meets the requirements.

(5) t (s) represents the running time of the algorithm. If the running time is not very long, it can be used as an offline algorithm.

6.4 A Case Study of the Qinghai–Tibet Railway Line

6.4.1 Data Declaration

The Qinghai–Tibet railway, a high-altitude railway linking Qinghai province's Golmud basin to Lhasa in the Tibet autonomous region, was completed and opened on July 1, 2006. Figure 6.1 shows the first train after the inauguration, and Fig. 6.2 shows the main stations along the Qinghai–Tibet railway and the mountains along the railway. The Qinghai–Tibet railway has created many new "first" in the history of railway construction. For example, Tanggula station on the Qinghai–Tibet railway line, which is 5072 m above sea level, is the highest railway station in the world. The wind-volcano tunnel is 4095 m above sea level and is the highest train tunnel in the world.

By using differential GPS technology, 500,000 data points with accuracy up to cm are obtained from Qinghai–Tibet railway. The distance between the adjacent data points is 1.5 to 3 m, and the average distance is about 2.5 m. The GPS data in Tanggula section includes GPS data that starts at Tanggula station and extends in opposite directions. This part of the data is used as a verification data set to test the performance of the algorithm.

Fig. 6.1 Train on the QTR

Fig. 6.2 The main stations and the mountains along Qinghai–Tibet railway

Firstly, the GPS data was placed in the three-dimensional coordinates with longitude, latitude, and altitude as the coordinate axes. Secondly, the starting point of the railway section serves as the origin point of coordinates. The railway section selected is 24 km long, including 9587 data points. Under the selected origin point, we assume that the direction from the origin station to Golmud is positive, so the direction to Lhasa is negative. According to the definition of this direction, we divided the data set into two parts, taking Tanggula station as the origin point, with 4622 data points in the positive direction and 4966 data points in the opposite direction.

6.4.2 Algorithm Results Analysis and Comparison

Before comparing the three algorithms, we first give some results of the projection calculation. Data reduction is to create directed graph by 2δ method or to find the shortest path by V method. 2δ method is used to reduce calculation time, save storage space, and make the error constraint is limited to the range of 2δ. The proof process is illustrated in Sect. 6.4.3.

Table 6.1 shows that using 2δ method can effectively reduce projection calculation. Applying 2δ method alone can effectively reduce calculation. The calculation quantity of data set 1 and data set 2 can be decreased by up to 95 and 85%, respectively. Using the V method alone, the data set 1 and data set 2 roughly reduce the amount of calculation by about 50%. To sum up, these two methods can greatly reduce the calculation time of Algorithm 3 and improve the operation efficiency. Performance comparisons of three algorithms on data set 1 and data set 2 are shown in Tables 6.2 and 6.3, respectively.

(1) It can be seen from Table 6.3 that the values of r for all three algorithms are close to 1, indicating that all three algorithms are effective for GPS data reduction. In other words, there are too many redundant data points in the collected GPS data. In terms of reduction efficiency, Algorithm 3 is the best, Algorithm 2 is the worst.
(2) In terms of running time, for data set 1, Algorithm 3 is the slowest and Algorithm 2 is slightly faster than Algorithm 1. Algorithm 3 is faster on data set 2 because its corresponding lines contain many straight lines.
(3) Because the data set 1 contains more curve lines than the line corresponding to the data set 2, the r value corresponding to the data set 1 is less than that corresponding to Algorithm 2 within the same exact error range δ. With the increase of error coefficient, the r value of the three algorithms will increase accordingly.

Table 6.1 Projection calculation comparison with or without 2δ method and V method

Projection calculation	Without 2δ method		2δ method			
			Without V method		With V method	
	$\delta = 0.5$	$\delta = 1$	$\delta = 0.5$	$\delta = 1$	$\delta = 0.5$	$\delta = 1$
Data set 1	1.6446e+10	1.6446e+10	2.1402e+08	3.1271e+08	0.8638e+08	1.4150e+08
Data set 2	2.0399e+10	2.0399e+10	1.5130e+09	2.6602e+09	8.0326e+08	8.1796e+08

Table 6.2 Performance comparison of three algorithms on data set 1

δ	0.5 m			1 m		
Algorithm type	1	2	3	1	2	3
r(%)	97.86	96.97	97.94	98.53	98.05	98.55
e_l(‱)	0.79	0.39	0.77	1.60	0.99	1.61
e_a(m)	0.30	0.15	0.31	0.57	0.34	0.61
e_m(m)	0.50	0.50	0.50	1.00	0.97	1.00
t(s)	50.5	16.4	3684	63.5	13.7	5998

(4) All three algorithms have very small e_l values. For the selected station, if the error is 0.5 or 1 m, the length loss is 1 or 3 m, respectively. Therefore, it is feasible to approximate the path of the railway line by using the straight line.
(5) The e_m and e_a values from the table show that all three algorithms satisfy the constraint Eq. (6.4). Algorithms 1 and 3 make full use of the constraint conditions, but Algorithm 2 does not.

Due to space limitations, only some calculation results diagrams were selected for demonstration in this chapter. Since the elevation change of the two data sets is very small, only two-dimensional calculation results are shown. The points in Figs. 6.3, 6.4, 6.5 and 6.6 represent the key points of GPS data discovered by the three algorithms. The three algorithms shown in Fig. 6.7 have been put into practical engineering application after many actual tests.

Table 6.3 Performance comparison of three algorithms on data set 2

δ	0.5 m			1 m		
Algorithm type	1	2	3	1	2	3
$r(\%)$	99.54	99.34	99.58	99.68	99.60	99.70
$e_l(‱)$	0.092	0.051	0.092	0.18	0.098	0.17
e_a(m)	0.29	0.13	0.29	0.56	0.33	0.49
e_m(m)	0.50	0.45	0.50	1.00	0.87	1.00
t(s)	149.4	16.1	23,131	253.5	13.9	23,532

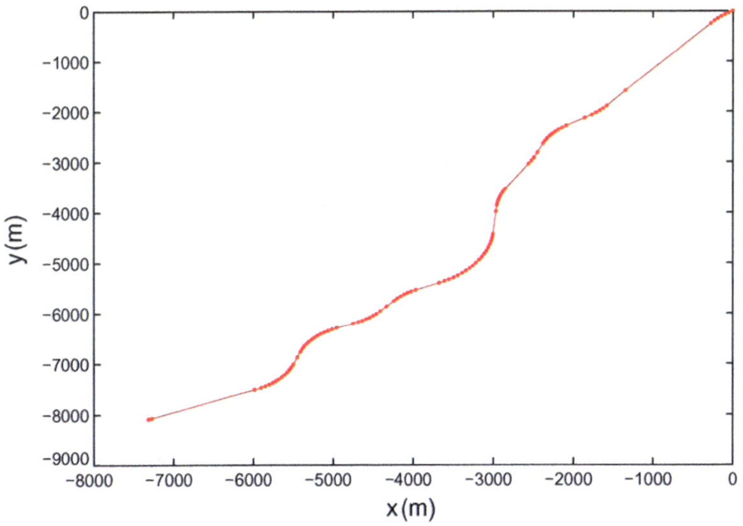

Fig. 6.3 Key points in data set 1 found by Algorithm 1 for a 0.5 m error bound

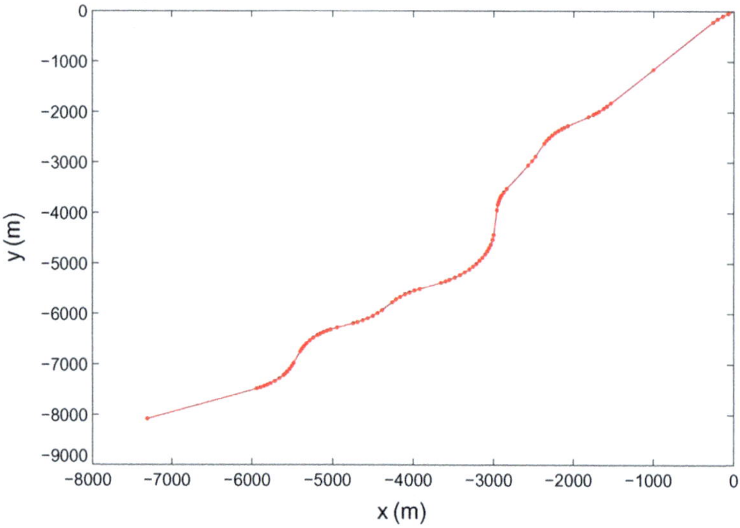

Fig. 6.4 Key points in data set 1 found by Algorithm 2 for a 1 m error bound

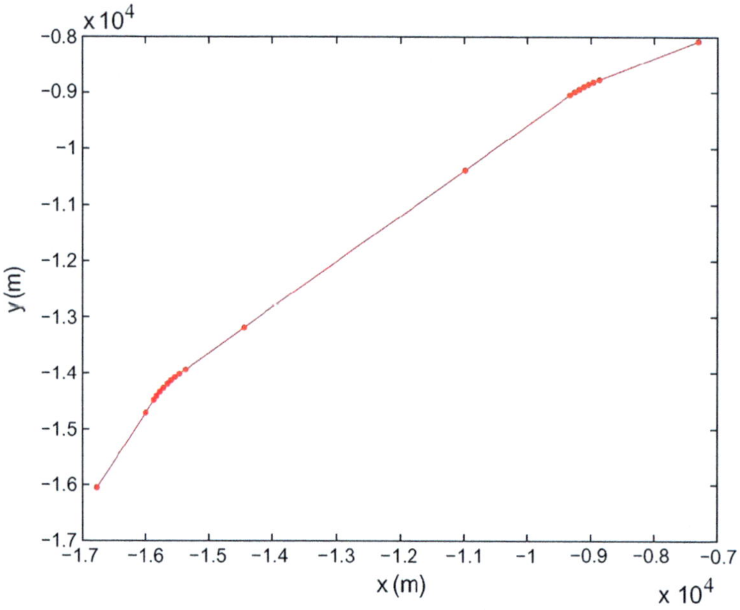

Fig. 6.5 Key points in data set 2 found by Algorithm 3 for a 0.5 m error bound

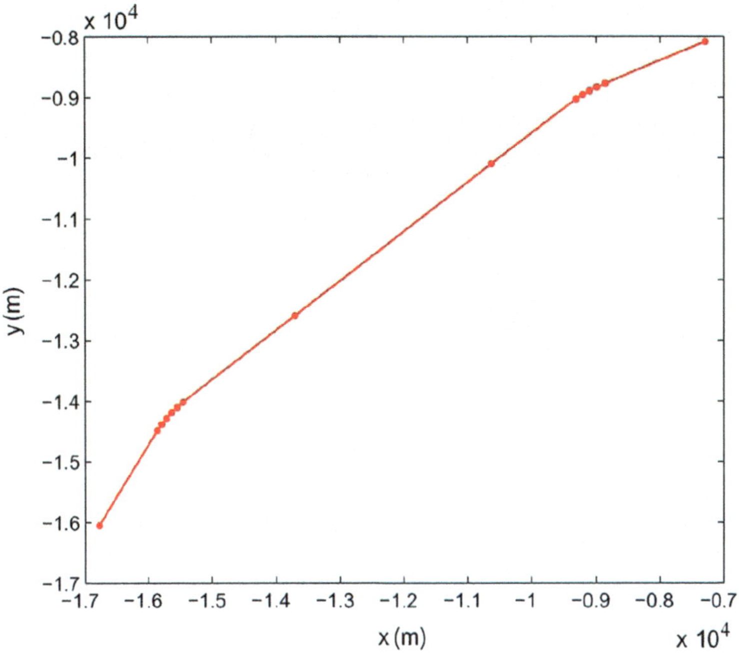

Fig. 6.6 Key points in data set 2 found by Algorithm 1 for a 1 m error bound

6.4.3 *Proof of 2δ Method*

On the premise of assumption 1.1, the following lemma 1.2 can be proved. The lemma shows that no feasible line segments are missed when an adjacency matrix is produced by the 2δ method.

Assumption 1.1: if the $\overline{p_i p_h}$ meets the 2δ condition for all $h = i + 1, \ldots, j$, that is

$$\left\| p_{j+1} - p_i \right\|_2 \geq \| p_h - p_i \|_2 + \delta \tag{6.11}$$

If the given data points are in the same order, the assumption is reasonable when the δ is assumed small. If all GPS data points from i to j are approximately on a straight line (within 2 m error), the distance of the next GPS data point is at least equal to δ. Corresponding to the rail, that is, if the given railway section from the start point p_i to the end point p_j approximates on a straight line, the train usually passes through a GPS data point p_{j+1} (from p_j to p_{j+1}) in the same direction, i.e., the distance between p_j and p_{j+1} is at least δ.

Lemma 1.2 In the case of assumption 1.1, if $\overline{p_i p_j}$ does not meet the constraint of 2δ, then for any $l \geq j + 1$, $\overline{p_i p_l}$ also do not meet the δ conditions, i.e., for all $l \geq j + 1$, $A(i, l) = 0$.

Fig. 6.7 Field test for algorithms on the Qinghai-Tibet Railway

Proof If $\overline{p_i p_j}$ does not meet the conditions of 2δ, there will exist p_k meets Eq. (6.12),

$$\text{dist}(L(i, j+1), k) > 2\delta, i < k < j+1 \tag{6.12}$$

In general, we assume that for all $l = i+1, \ldots, j$, D(i, l) $\leq 2\delta$. Define the data set $S_1 = \left\{ \begin{array}{l} p: \text{the projection distance from } p_{j+1} \text{to } \overline{p_i p} \text{ is less than or equal to the } \delta, \\ \text{and } p_{j+1} - p_i, p - p_i \geq 0 \end{array} \right\}$

$$S_0 = \{Q : Q = p - p_i, p \in S_1\} = S_1 - \{p_i\} \tag{6.13}$$

Here, S_0 is convex cone and defines $S_2 = \{p_i\} - S_0$, $S = S_1 \cup S_2$. Calculate the projection of point p_k to data set S and line $\overline{p_i p_j}$, and obtain the projection points Q_1 and Q_2. Then, calculate the projection of point Q_1 to line $\overline{p_i p_{j+1}}$, and obtain the projection points Q_3. Calculate the projection of point p_{j+1} to line $\overline{p_i Q_1}$, and obtain the projection points Q_4. Calculate the projection of point Q_4 to line $\overline{p_i p_{j+1}}$, and obtain the projection points Q_5.

Because Q_2 is the projection from p_k to $\overline{p_i p_{j+1}}$, Eq. (6.14) can be obtained based on assumption 1.1:

$$\|p_i - Q_2\| \leq \|p_i - p_k\| \leq \left\|p_i - p_{j+1}\right\| - \delta \tag{6.14}$$

Similarly, we can obtain Eqs. (6.15) and (6.16).

$$\|p_i - Q_1\| \leq \|p_i - p_k\| \leq \left\|p_i - p_{j+1}\right\| - \delta \tag{6.15}$$

$$\|p_i - Q_3\| \leq \|p_i - p_k\| \leq \left\|p_i - p_{j+1}\right\| - \delta \tag{6.16}$$

For Q_5 is the projection from Q_4 to $\overline{p_i p_{j+1}}$, point Q_5 must be on sphere $B(p_{j+1}, \delta)$ and $\left\|Q_5 - p_{j+1}\right\| < \delta$. So, $\|p_i - Q_5\| > \left\|p_i - p_{j+1}\right\| - \delta$, for which Eqs. (6.17) and (6.18) can be obtained.

$$\|p_i - Q_3\| \leq \|p_i - Q_5\| \tag{6.17}$$

$$\|Q_1 - Q_3\| < \|Q_4 - Q_5\| < \left\|p_{j+1} - Q_4\right\| = \delta \tag{6.18}$$

Because $\|p_k - Q_2\| < \|p_k - Q_1\| + \|Q_1 - Q_3\|$ and $\|p_k - Q_2\| > 2\delta$, $\|p_k - Q_1\| > \delta$. Suppose there is a point $p_l(l > j+1)$ that makes $\overline{p_i p_l}$ satisfy δ, then p_l must be in the data set S. However, because the projection from p_k to S is bigger than δ, $\text{dist}(L(i,l), j+1)$ and $\text{dist}(L(i,l), k)$ cannot both be less than or equal to δ. So, there is no point $p_l(l > j+1)$ that makes $\overline{p_i p_l}$ satisfy δ.

6.5 Brief Summary

The problem of GPS data reduction of Qinghai–Tibet railway is a complex combinatorial mathematical problem. In this chapter, two heuristic algorithms and an optimization algorithm are firstly proposed, and the performance of the algorithm is verified by using the measured data. The experimental results show that the performance of the three algorithms is good. Data reduction is to create directed graph by 2δ method or to find the shortest path by V method.

The results show that the heuristic algorithm runs faster than the optimization algorithm. Taking both reduction efficiency and running time into account, we suggest using Algorithm 1 to calculate large data sets. Algorithm 2 is used to verify the existence of shape data corresponding to Z glyph. Therefore, Algorithm 1 is an ideal heuristic algorithm for railway GPS data reduction.

Usually, in practice, the railway GPS data reduction problem requires larger error bounds, straighter track, and higher data compression rate. Therefore, the data reduction algorithm proposed in this chapter will effectively save storage space and accelerate the speed of train positioning. In addition, the real-time train positioning algorithm based on data compression needs further research.

References

1. Blomenhofer H. GNSS in the 21st century-the user perspective. Acta Astronaut. 2004;54:965–8.
2. Skog I, Handel P. In-car positioning and navigation technologies—a survey. IEEE Trans Intell Transp Syst. 2009;10(1):4–21.
3. Zhang J, Chen D, Kruger U. Adaptive constraint K-segment principal curves for intelligent transportation systems. IEEE Trans Intell Transp Syst. 2008;9(4):666–77.
4. Santos AJD, Soares AR, de Almeida Redondo F M, et al. Tracking trains via radio frequency systems. IEEE Trans Intell Transp Syst. 2005;6(2):244–58.
5. Raymond G, Marais J, Berbineau M. Innovation brings satellite-based train control within reach. Railway Gaz Int. 2004;160(12):835–7.
6. Urech A, Pérez Diestro J, Gonzalez O. GADEROS, a GALILEO demonstrator for railway operation system. In Data systems in aerospace, Dublin, Ireland; 2002. p 442–447.
7. Albanese A, Marradi L, Labbiento G, et al. The RUNE project: the integrity performances of GNSS-based railway user navigation equipment. In Proceeding of ASME/IEEE 2005 joint rail conference. American Society of Mechanical Engineers; 2005. p. 211–8.
8. Baker J, Clennan J. ITCS: a new approach to increasing capacity. Int Railway J. 2005;45(10):46–7.
9. Li K. The project of ITCS signal control system of Qinghai-Tibet Railway. Chin Railways. 2005;7:31–6.
10. Simsky A, Wilms F. GNSS-based failsafe train positioning system for low-density traffic lines based on one-dimensional positioning algorithm. Conserv. Matters. 2004.
11. Gu YH, Tjahjadi T. Coarse-to-fine planar object identification using invariant curve features and B-spline modeling. Pattern Recogn. 2000;33(9):1411–22.
12. Gao G, Cai B. Research on the automatic electronic map generation algorithm for the train supervision system. J China Railway Soc. 2006;1:63–70.
13. Bruald RA. Introductory combinatorics. 4th ed. Englewood Cliffs, NJ: Prentice-Hall; 2004.
14. Pemmaraju S, Skiena S. Computational discrete mathematics: combinatorics and graph theory with mathematica. Cambridge, U.K.: Cambridge University Press; 2003.

Chapter 7
Multiple GPS Track Information Fusion

7.1 Overview

7.1.1 GPS Digital Map

GPS is a new generation space satellite navigation and positioning system developed by the US army, navy, and air force in the 1970s. Its main purpose is to provide real-time, all-weather, and global navigation services in land, sea, and air. GPS consists of 24 working satellites, which are located 20–200 km above the earth's surface and distributed evenly on six orbital planes (four for each orbital plane), with an orbital inclination of 55°. In addition, there are three active backup satellites in orbit. The distribution of satellites allows more than four satellites to be observed anywhere in the world at any time and can store navigation information in the satellite.

With the accelerated pace of urban construction and renovation in China, urban roads are constantly being built, and complex road traffic has even spawned a number of operators. New buildings, new asphalt roads make some familiar places strange, and crisscrossing urban roads and different types of overpasses make the residents of the city have to check the map frequently when they go out. On the one hand, the booming sales of maps reflect the rapid development of the city; on the other hand, they also reflect people's pursuit of fast-paced life. Convenient and accurate arrival to the destination can not only save precious time, but also save high fuel costs for drivers. So, young people familiar with the Internet are beginning to use digital maps.

Digital map is the digitized appearance of paper map, is the discrete data of ground elements and phenomena with definite coordinates and attributes in a certain coordinate system, and is a generalized and orderly set on the identifiable storage medium of computer. A digital map based on a map database is stored in digital form on an external computer reservoir and can be displayed on an electronic screen. In high development of computer technology and information science, only paper maps and some scattered digital maps provide information has been unable to meet this need. Instead of paper maps, based on the regional or national digital maps, all

D. Chen and R. Cheng, *Intelligent Processing Algorithms and Applications for GPS Positioning Data of Qinghai-Tibet Railway*, https://doi.org/10.1007/978-3-662-58970-0_7

kinds of map database management system and geographical information system are applied in the plane, the ships' navigation, the new weapon guidance, satellite and military rapid response, military command automation, and the various industries of economic construction. Digital map can is very convenient for combining and matching the content of general map in any form to form a new map. Since digital maps can be output at any scale and range, it is easy to modify, which can greatly reduce the time for making maps; it can be easily combined with other information sources such as satellite images and aerial photographs to generate new types of images. At the same time, we can make use of the recorded information of digital maps to derive new data, such as the contour lines representing topographic features on the map. However, using the contour lines is very difficult for non-professional personnel to understand. The contour and elevation point of digital maps can be used to generate digital elevation model and display the contour in digital form, which can be intuitive to the three-dimensional geomorphic feature. This is an effect that ordinary topographic maps cannot achieve.

At the same time, accurate railway digital map plays a very important role in train operation, safety, accurate positioning, train scheduling, and many other aspects in train operation control system [1]. It is a kind of navigation information source that is not affected by the operational environment and has high stability, which enables the train know all kinds of information about the line. If digital track map is involved in train positioning, it can effectively correct errors and improve positioning reliability and accuracy [2]. Similarly, accurate road digital maps play an important role in vehicle navigation, route selection, and vehicle induction.

7.1.2 Principal Curve

Spearman proposed the linear principal component analysis (PCA) method in 1904, which is simple and easy to use and is still an important tool for data statistical analysis [1]. The principle of linear PCA is to project the data set to a vector so that the mean square deviation of the data set to the projection vector is minimal. And, this vector is called the first principal component of the data set. However, the information of original data, such as cosine signal, may be lost by using principal component analysis. Therefore, Trevor Hastie of the department of statistics of Stanford University proposed the concept of the main curve in 1983 [2]. The principal curve is similar to the linear regression method in the non-linear extension, instead of linear principal component to summarize data, using smooth curve to find the curve between the symmetric variables. Such a curve passes smoothly from the middle of the data and is not limited to the smooth linear average of the data, or even the middle of the data is a straight line, only in the hope of minimizing the orthogonal distance from the data point set to the curve.

Principal curves are widely used, such as the control of electron beam trajectory in linear collider [3], the recognition of ice sheet contour [4] in image processing, the principal curve template of handwriting [5–7], the simplified modeling and audibility of sound data in speech recognition [8–10], the orderly distribution of population in ecology [11], and process monitoring [12, 13]. At the same time, in the field of cognition, Seung proposed that perception exists in manifold mode [14] and found that the trigger rate of the whole nerve cell population in neurophysiology can be described by a function composed of a few variables, such as the angle of the eye [15] and the direction of the head [16].

In 1992, Banfield and Raftery principal curves (abbreviated as BR principal curves) improved the algorithm of principal curves and solved the problem that the estimation deviation had a significant impact on the actual principal curves under closed principal curves due to the excessive curvature [4]. However, the BR principal curve method also introduces numerical instability, which may produce smooth but incorrect principal curves in practice. In 1994 and 1995, Duchamp and Stuezle investigated integral differential geometry features of principal curves in [17, 18], analyzing the relationship between self-consistency and curvature, and the relationship between first variation of the principal curve and the second variation of the principal curve. They pointed out that when the precondition of curvature is not zero, the principal curve in the plane is the saddle points with the minimal expected square distance to the data, and there is no local minimum. For this reason, using an evaluation of expected square distance (such as cross-validation) cannot be used to select the model complexity method of the principal curve. This study also illustrates why the principal n curve is not unique.

In 1997, Kegl introduced the concept of the principal curve with length constraint [5–7], which proved the existence and uniqueness of K principal curve defined under the theoretical distribution. The K principal curve must exist as long as there is a finite second moment in the data distribution. At the same time, the learning property of K principal curve for data distribution is studied, the convergence proof and convergence rate of newly defined K principal curve are given, and the corresponding polygon line algorithm is proposed to find K principal curve by combining with the complexity thought in statistical learning theory. This is the first paper to prove the existence of the principal curve. The defect is that the length of the curve must be fixed in advance. In 2000, Verbeek proposed the soft k-segment principal curve algorithm [19], which used the local principal component method to generate K line segments and connected them to form the principal curve according to the smoothness. The defect of this method is that it cannot be extended to higher dimensions, and the curvature penalty is not considered in the objective function.

In order to solve the practical application problem of the principal curve, Junping Zhang and Dewang Chen proposed the constraint principal curve algorithm [20] in 2006. The algorithm solved the generation and adaptability problems of the principal curve with constraint points, but the calculation method was too complicated and the practicability was not high. In 2008, Dewang Chen proposed a heuristic optimization algorithm for multiple track information fusion based on principal curve theory [21], but there are still problems such as long operation time and narrow application range.

In this chapter, based on the theory of the existing K principal curve, the generation and optimization methods of principal curve algorithm are made further research and improvement. Then, we proposed constraints principal curve algorithm, which is practical in engineering practice. Three different methods are proposed for the optimization of principal curves: the splitting algorithm with increasing vertices, the improved split optimization algorithm carried out step by step, partition and merge optimization algorithm which divides the region and then merges.

Then, the algorithm was developed by using C# and matlab, and the calculation error, stability and fitness of complex trajectory were studied through the simulation experiment of simulated data. The practical aspects of the algorithm, such as data storage space and computational efficiency, are studied and analyzed through the measured GPS data of railway and highway. The results show that the practical constrained principle curve algorithm proposed in this chapter has good performance in calculation accuracy, calculation efficiency, data storage, complex track fitness, and algorithm stability.

7.2 The Mathematical Model of K Principal Curve Algorithm

The constraint principal curve is established on the basis of the model of k-segment principal curve, and its main goal is to use the nearest neighbor principle to find a principal curve composed of polyline across the center of the data set. Usually, they are obtained by gradually increasing the fixed point, and the vertex of the principal curve needs to satisfy some constraints (such as fixed endpoint), as shown in Fig. 7.1. Constraint principal curve is made up of the line $S_{j,j+1}(j = 0, 1, 2, \ldots, n-1)$ between vertex $V_j(j = 0, 1, 2, \ldots, n)$. The vertex positions of V_0 and V_n are fixed.

The nearest neighbor principle is to divide the data point according to the projection from each data point to all line segment and vertex of the principal curve. In other words, it is the rules that the data point $d_i(i = 0, 1, 2, \ldots, m)$ is divided into the nearest line segment or vertex region on the principal curve. n is the number of vertex vectors and m is the number of GPS data points.

The distance from the data point to the corresponding line is calculated by Eq. (7.1).

$$\mathrm{DL}_{i,j} = \frac{\left|\left(X_{\mathrm{v}_j} - X_{\mathrm{v}_{j+1}}\right) \cdot Y_{\mathrm{d}_j} - \left(Y_{\mathrm{v}_j} - Y_{\mathrm{v}_{j+1}}\right) \cdot X_{\mathrm{d}_j} + X_{\mathrm{v}_{j+1}} \cdot Y_{\mathrm{v}_j} - Y_{\mathrm{v}_{j+1}} \cdot X_{\mathrm{v}_j}\right|}{\sqrt{\left(X_{\mathrm{v}_j} - X_{\mathrm{v}_{j+1}}\right)^2 + \left(Y_{\mathrm{v}_j} - Y_{\mathrm{v}_{j+1}}\right)^2}} \tag{7.1}$$

where X_{d_j} and Y_{d_j} are the ordinates of data point d_i, X_{v_j} and Y_{v_j} are the ordinates of vertex v_j, segment $S_{j,j+1}$ link v_j and v_{j+1}.

The distance between the data point to the vertex is calculated by Eq. (7.2).

$$\mathrm{DP}_{i,j} = \sqrt{\left(X_{\mathrm{d}_i} - X_{\mathrm{v}_j}\right)^2 + \left(Y_{\mathrm{d}_i} - Y_{\mathrm{v}_j}\right)^2} \tag{7.2}$$

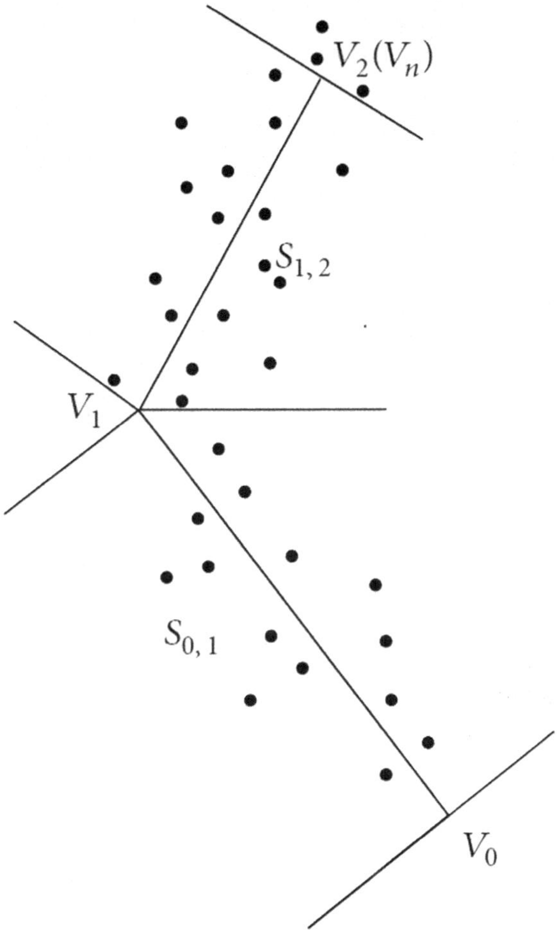

Fig. 7.1 Principle of nearest neighbor

The minimum of Distance $\mathrm{DL}_{i,j}(j = 0, 1, 2, \ldots, n)$ and distance $\mathrm{DP}_{i,j}(j = 0, 1, 2, \ldots, n)$ is defined as the projection distance of K section principal curve. The average projection distance of all data points AD is defined as follows, that is, the vertical distance from each data point to the nearest vertex or line segment on the target principal curve is defined as the projection distance.

$$\mathrm{AD} = \frac{\sum_{i=1}^{m} \min\{\mathrm{DL}_{i,j}, \mathrm{DP}_{i,j}\}}{m}, \quad j = 0, 1, 2, \ldots, n \tag{7.3}$$

In order to minimize AD, the number and location of line segments should be optimized. However, since we do not know how many line segments in advance, it is not easy to get the optimal solution in one step. Therefore, in practical engineering application, in order to avoid using the complex and long-term algorithm, we use the iterative algorithm to solve the problem.

7.3 Three Iterative Algorithms Based on Constraint K Principal Curve

According to the above mathematical model, we propose three iterative algorithms to generate constraint K principal curves in this section. For the three algorithms, two vertices of the curve (V_0 and V_n) are served as constraint points and remain unchanged during the iteration. But different algorithms mean using different methods to find the final solution, adding some other vertices according to region knowledge. Many successful applications show that good algorithms and intelligent efficient algorithms are implemented with the help of expert knowledge [17–19].

For three algorithms, it is a traditional non-linear optimization problem to minimize the objective function by adjusting the position of several or all vertices when the number of vertices is fixed. This problem has been reflected in the non-linear optimization toolbox function "fmincon" in matlab program. The "fmincon" function uses traditional SQP [20] to optimize the small range problem and the large range problem of confidence interval [19].

7.3.1 The Splitting Algorithm with Increasing Vertices

In this algorithm, two main steps are needed to obtain the constraint K principal curve: dividing the data into blocks and optimizing all the vertices. The algorithm is referred to as ALLopt algorithm. The main steps of the algorithm are as follows:

Step 1: Connect the two endpoints as the initial target principal curve.
Step 2: Calculate whether the average error AD of all data points from d_i to the initial target principal curve is less than the preset error E. If it is satisfied, then the principal curve generation ends. If it is not satisfied, jump to Step 3.
Step 3: Add the vertex v_j between the two endpoints v_0 and v_n to form the new target principal curve. The position of the increasing vertex is the last principal curve segment. Use the non-linear optimization method to adjust the position of all vertices (among which the position of the constraint point of the principal curve cannot be moved) to minimize the average error AD.
Step 4: Calculate the average error AD of all data points from d_i to the target principal curve and determine whether AD is less than the preset error E. If not, repeat Step 3, continue to increase the vertex in the last section of the principal curve segment, and then optimize until it is satisfied. The flow chart of the split optimization algorithm is shown in Fig. 7.2.

Fig. 7.2 The flow chart of the split optimization (ALLopt) algorithm

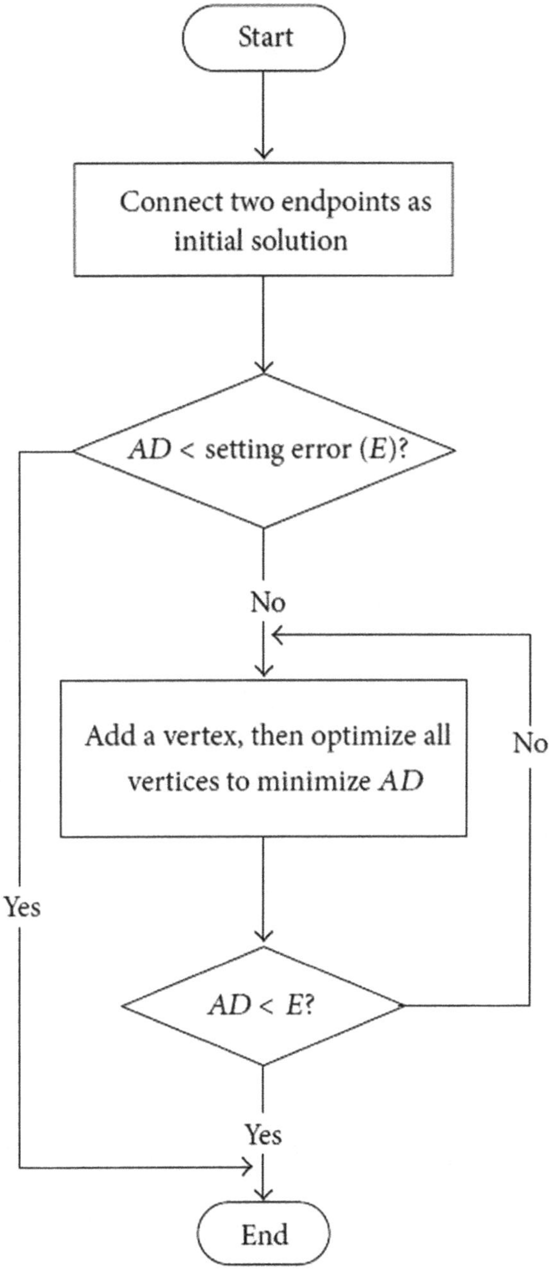

7.3.2 The Improved Split Optimization Algorithm-MPMopt

Because the split optimization algorithm optimizes all the vertices at each iteration, it takes a long time to run the program with a large amount of data. In order to solve this problem, we propose an improved algorithm that applies the initial solution well and adjusts three vertices at most once to speed up the process of optimization. The algorithm consists of two parts, the first part is to generate an initial principal curve, and the second part is to calculate and solve the final principal master curve. The algorithm is referred to as MPMopt algorithm.

Part one: Generate the initial principal curve:
Step 1: Connect the two endpoints as the initial target principal curve;
Step 2: Calculate whether the average error AD from the data point d_i to the initial target principal curve is less than the preset error E. If it is satisfied, then the principal curve generation ends. If it is not satisfied, jump to Step 3.
Step 3: Increase the vertex v_j between two endpoints v_0 and v_n to form a new target principal curve and calculate whether the average error AD from all data points d_i to the target principal curve is less than the k times of preset error E (k is usually 2–4). If not, repeat Step 3 and continue to increase the vertex; if the initial principal curve is satisfied, the generation ends.
The location selection method for adding new vertices is as follows: Select the region with the maximum average error in the linear region of each segment of the target principal curve, and then select the point with the maximum distance to the corresponding vertex and line segment of the target principal curve in this region. Finally, calculate the average coordinates of the data points of the neighborhood of the selected point as the coordinates of the added points. We call this method here Max Point Method (MPM).

Part two: Solve the final principal curve.
Step 4: On the basis of the target principal curve obtained from the generation of the initial trajectory, select the region with the maximum average error in each linear region of the target principal curve; add the initial position of a vertex v_j' in this region, the initial position of v_j' is the mean coordinates of data points in the neighborhood of the data point with the maximum distance to corresponding vertex and line segment from the target principal curve.
Step 5: Locally optimize the position of new vertex v_j' and two adjacent vertices v_j, v_{j+1}, so as to minimize the average error (the position of constraint point remains unchanged); calculate the average error AD of the data point d_i to the target principal curve and determine whether AD is less than the preset error E. If not, repeat Step 4 and Step 5 to continue to increase the vertex; If it does, the principal curve ends.
The flow chart of the improved split optimization algorithm is shown in Fig. 7.3.

Fig. 7.3 The flow chart of the improved split optimization (MPMopt) algorithm

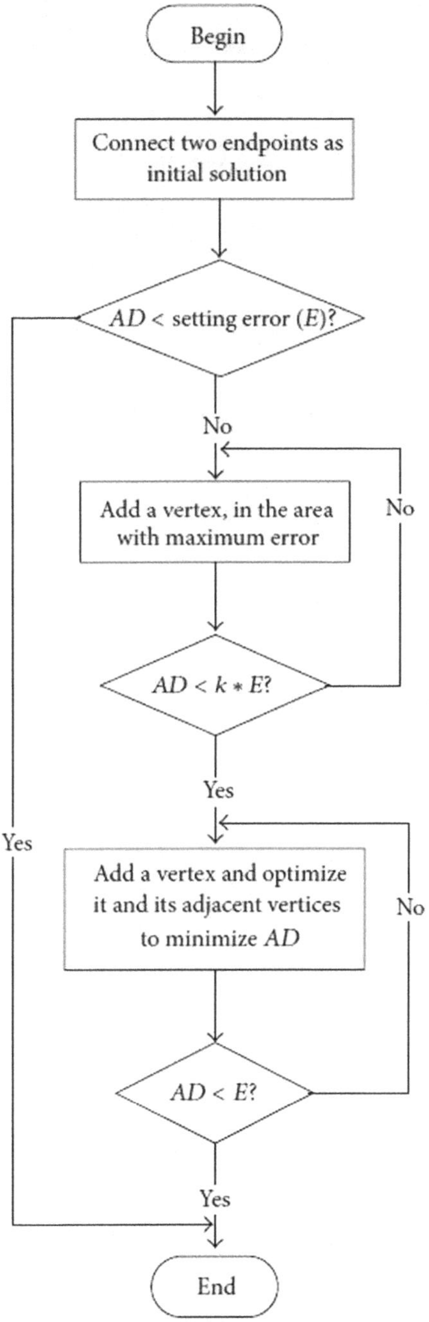

7.3.3 Division and Combination Optimization Algorithm-DCopt

DCopt consists of three parts: partitioning, merging, and optimization:

(1) Partition part: All data points are divided into the neighborhood range of each vertex of the principal curve. The specific steps are as follows:

Step 1: Divide the data points into the neighborhood of two endpoints according to the distance between the data points and the endpoints. Determine whether all data points are included in the neighborhood radius set of each vertex. If yes, go to the merging part; if not, jump to Step2.
Step 2: Add a vertex to the area with largest distance from the point to the two endpoints, re-divide the data point, and determine whether all data points are included in the neighborhood radius set of each vertex. If yes, go to the merging part; if not, jump to Step 2 and keep increasing vertices until all data points are contained within the neighborhood radius of each vertex.

(2) Merging part: The generated vertices are connected in order to form the principal curve according to the shape of the trajectory. Two rules are followed when generating the principal curve:

- The change rate of slope should be as small as possible;
- Connect the point with the smallest distance.

When connecting, the starting point with the smallest distance is selected according to rule 2, and then connect the point that follows rule 2 under the condition that rule 1 is satisfied.

(3) Optimization part: Optimize the position of each vertex of the generated principal curve to minimize the distance from the data point to the principal curve; determine whether the error requirements are met. If not, continue to increase the vertex for optimization until the error requirements are met, the position of the increased vertex is similar with the split optimization algorithm.
The flow chart of the entire algorithm is shown in Fig. 7.4.
In this section, the simulated data is used to analyze the error, stability, and applicability of the three algorithms, and the practicability of the algorithm is verified by the actual measurement of GPS data of Qinghai–Tibet railway. Qinghai–Tibet railway is the first railway line with GPS navigation technology in China.

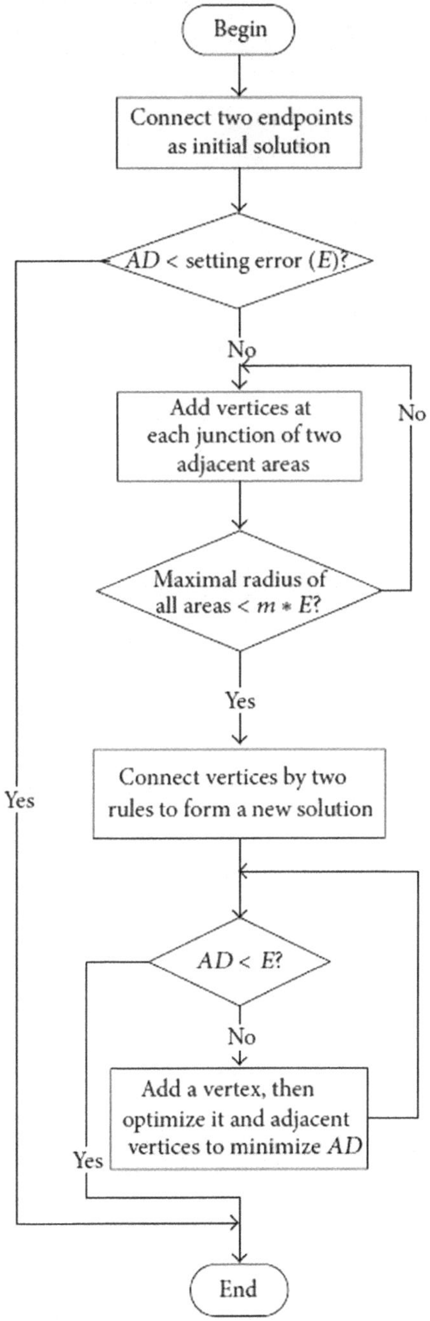

Fig. 7.4 The flow chart of DCopt algorithm

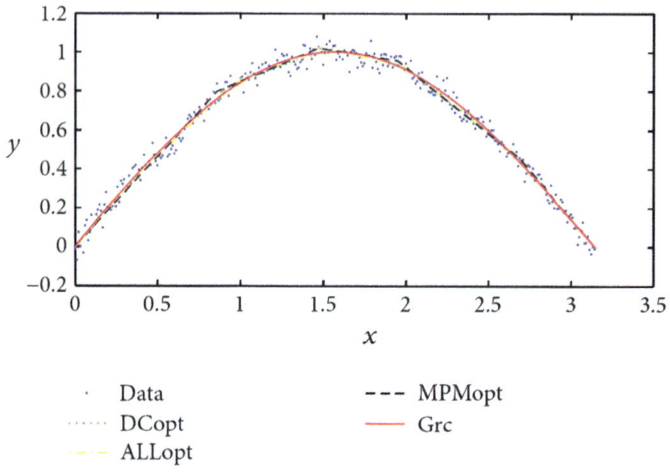

Fig. 7.5 Simulation error analysis of the three algorithms

7.3.4 Error Analysis

As shown in Fig. 7.5, three steps are required to synthesize the data. Firstly, generate a standard curve, and the two endpoints of the principal curve are fixed and known. Secondly, some points are randomly generated near the standard curve to simulate the real measurement data points. Finally, new data curves and data points are obtained through the above three algorithms, and the corresponding error results are obtained by comparing the new data curves with the standard data curves.

In Fig. 7.5, the red curve represents the generated reference curve $y = \sin(x)$, two fixed points are $(0, 0)$ and $(\pi, 0)$. The blue data points represent the generated random data points of the simulated measurement points. The curves in yellow, black, and green represent the curves generated by ALLopt, MPMopt, and DCopt, respectively.

To verify the performance of the measurement algorithms, each algorithm was simulated 30 times. For the reference curve, data points are randomly generated each time with a random error between 0 and 0.06. We found that all three algorithms can generate the principal curve well and make AD less than 0.03. Therefore, the three algorithms have better applicability and stability for simple curves.

7.3.5 Fitness Analysis

The simulation method is used to generate trajectory data of different shapes, including simple and complex trajectories. Then, the above three algorithms are used to generate the principal curve from the simulation data to test the fitness of the algorithm. The experimental results are shown in Figs. 7.6, 7.7, and 7.8.

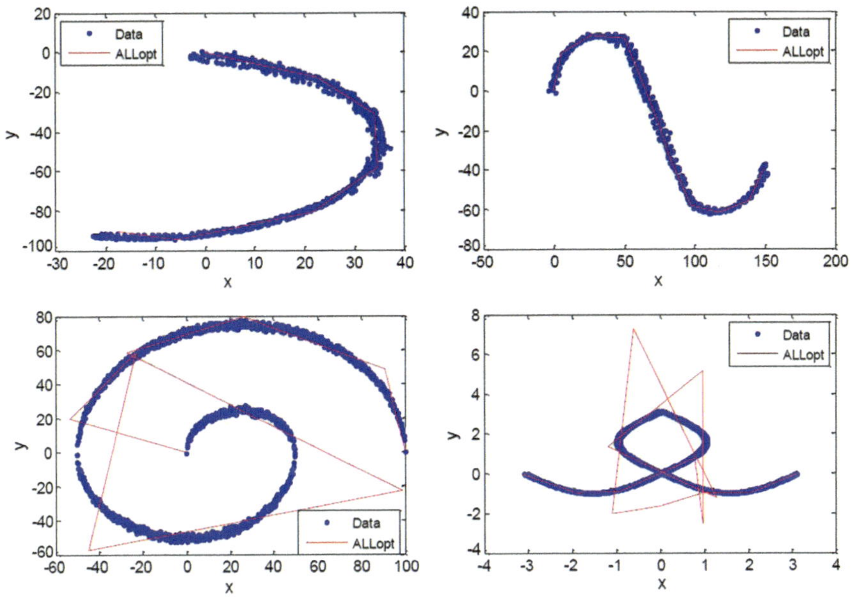

Fig. 7.6 Fitness analysis diagram of the split optimization algorithm

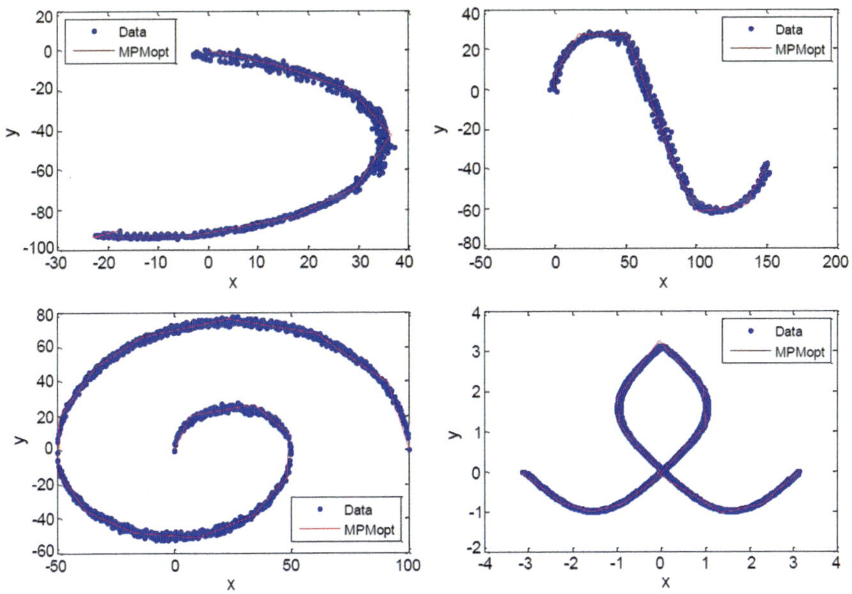

Fig. 7.7 Fitness analysis of the improved split optimization algorithm

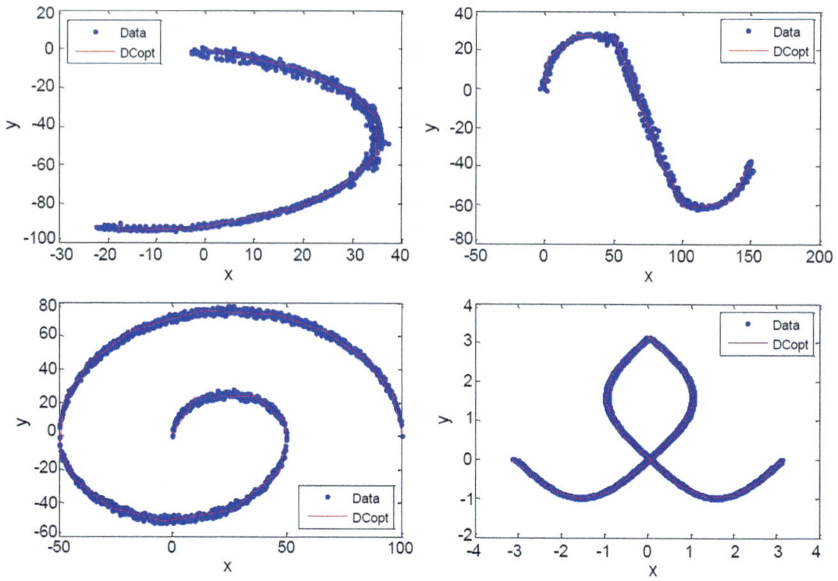

Fig. 7.8 Fitness analysis of division and combination optimization algorithm

It can be seen from Figs. 7.6, 7.7, and 7.8 that the split optimization algorithm has a good adaptability to common simple trajectory data, but it fails to achieve complex trajectories such as helix and cross. The improved split optimization algorithm has a good adaptability to common simple trajectory data and complex spiral trajectories, but it will cause problems to cross trajectories. Therefore, the improved split optimization algorithm is only applicable to part of complex trajectories. The division and combination optimization algorithms have good adaptability not only to ordinary simple trajectory data, but also to complex spiral or even cross trajectories, which can generate the main curve well.

7.4 Practical Application of the Algorithm

Railway has the following characteristics: simple line, long route, and large data volume. Therefore, we use measured GPS data of railway to test the practicability of the algorithm, and analyze the error, operation time, and storage space and other performance indexes of the algorithm. The railway GPS data is measured at *t*-04 and *t*-10 of the Nanqu logistics center in Tibet. The equipment used for the measurement is Thales *z*-max with an accuracy of 3 m.

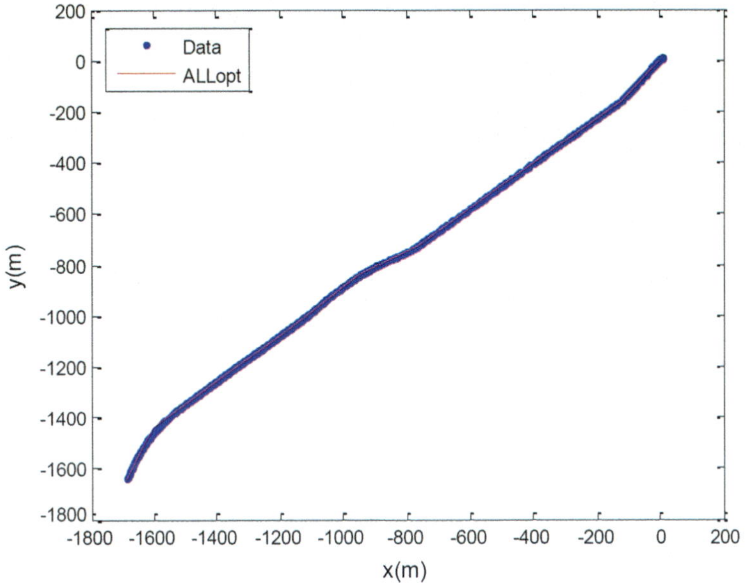

Fig. 7.9 The computation results of *t*-04 trajectory by using splitting optimization algorithm

t-04 is a railway track with a length of more than 2000 m in the station, and two terminals are two signal machines. The signal machines' position is fixed and needs to be measured accurately, so we measured the two endpoints for a long time of 30 min to get the exact coordinates of the endpoints. Then, three dynamic measurements were made on the trajectory, and 1093 data points were obtained. *t*-10 is a railway track with a total length of more than 40,000 m outside the station, and two endpoints are two switch turning machines. By the same measurement method, 15,213 data points are obtained.

7.4.1 Test Set 1

The above three algorithms are used to generate the principle curve by using the two measured GPS data of railways. The results of the GPS track of *t*-04 railways are shown in Figs. 7.9, 7.10, and Table 7.1.

It can be seen from Figs. 7.9, 7.10, 7.11, and Table 7.1 that all three algorithms can generate the principle curves well, but the time required by the split optimization algorithm is very long, while the time required by the improved split optimization algorithm is the shortest.

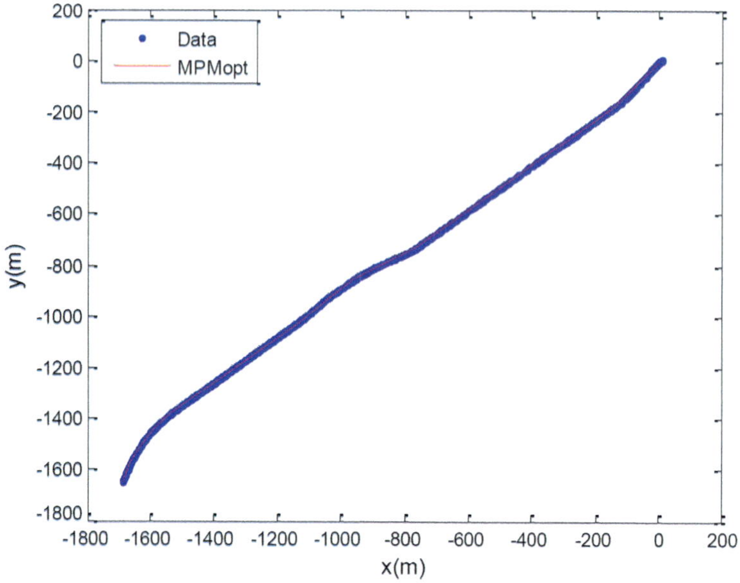

Fig. 7.10 The computation results of t-04 trajectory by using improved splitting optimization

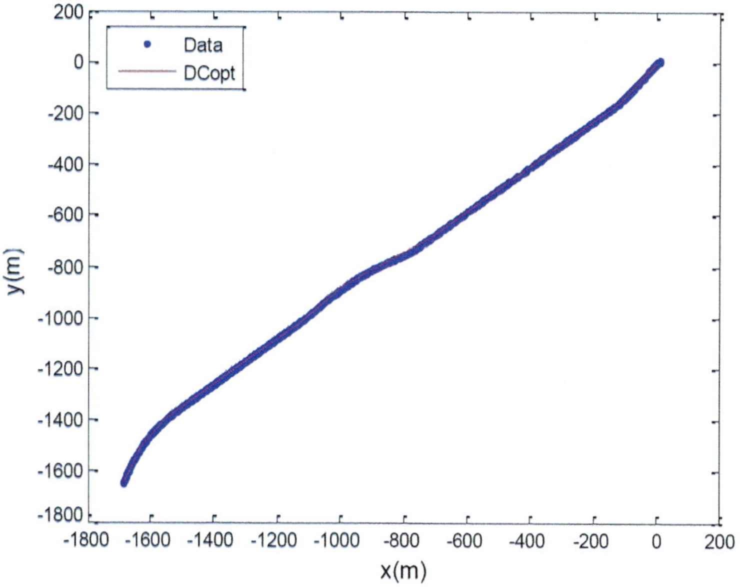

Fig. 7.11 The computation results of t-04 trajectory by using division and combination optimization algorithm

Table 7.1 The performance comparison of three algorithms for *t*-04 railway track

	Error (m)	Time (s)	Number
ALLopt	0.5572	72.8	8
MPMopt	0.5680	2.6	9
DCopt	0.6515	19.0	9

7.4.2 Test Set 2

The results of the GPS track of *t*-10 railways are shown in Figs. 7.12, 7.13, 7.14, and Table 7.2.

As shown in Figs. 7.12, 7.13, 7.14, and Table 7.2, the improved splitting optimization algorithm takes only 103 s for generating the principle curve, and the precision was also up to the requirements. The splitting optimization algorithm took 82 min and the error was up to 10 m. The division and combination optimization algorithm used 24 h to reach 27 m. It can be seen that when the trajectory is long and the data volume is large, the split optimization algorithm and the division and combination optimization algorithm fail due to the long calculation time. The improved split optimization algorithm can be completed in a very short time, and the storage points are greatly reduced. Only 68 points can be used to complete the storage of more than 15,000 data points.

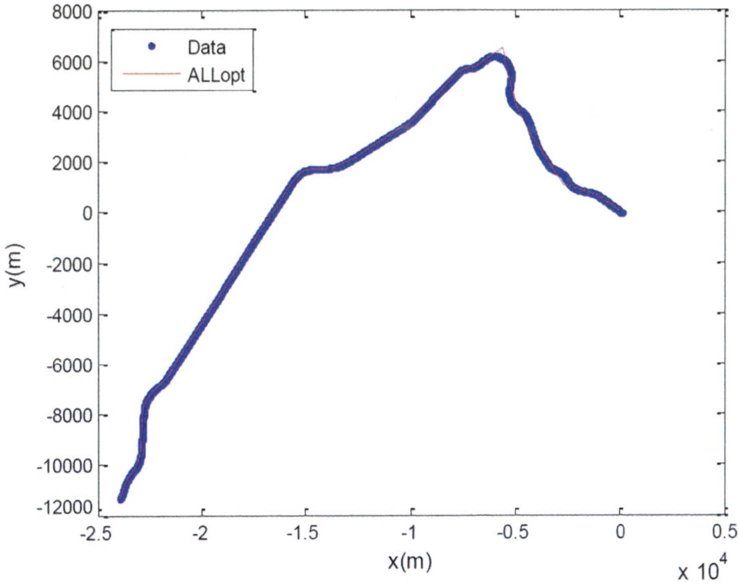

Fig. 7.12 The computation results of *t*-10 trajectory by using splitting optimization algorithm

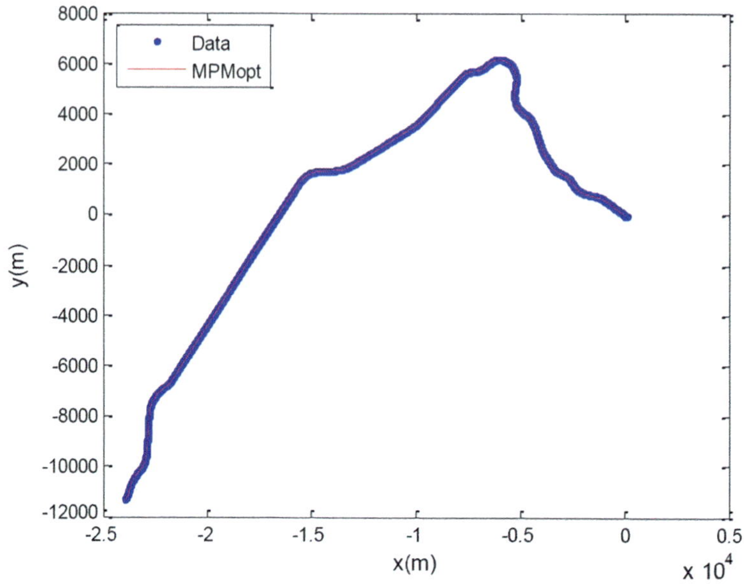

Fig. 7.13 The computation results of *t*-10 trajectory by using improved splitting optimization

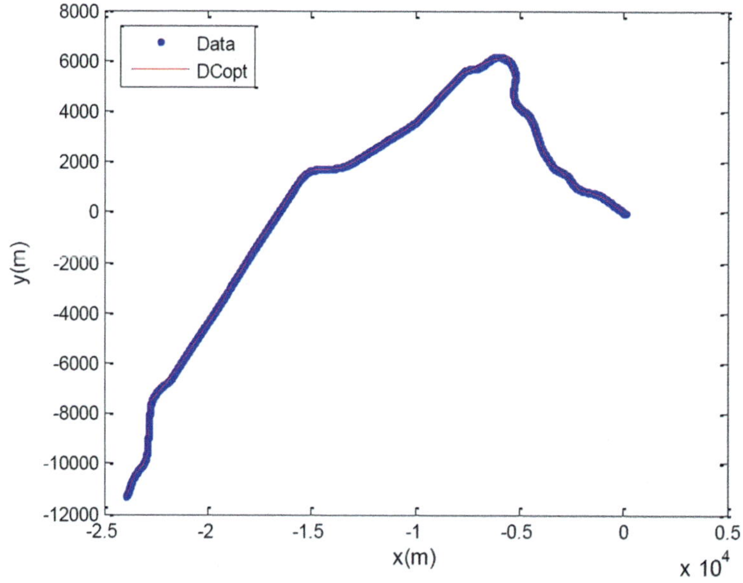

Fig. 7.14 The computation results of *t*-10 trajectory by using division and combination optimization algorithm

Table 7.2 The performance comparison of three algorithm for *t*-10 railway track

	Error (m)	Time (s)	Number
ALLopt	26.912	172832.2	56
MPMopt	0.9971	103.4528	68
DCopt	10.091	4940.191	51

7.5 Brief Summary

In this chapter, we first introduce the relevant knowledge, development history, and the deficiency in practical application of the theory of the principle curve. For solving the fixed endpoints problems, complex constraint k-segment principal curve is proposed. Then, in order to improve the practicability of complex algorithm, the principal curve algorithm is improved, and three practical constraint principal curve algorithms (ALLopt, MPMopt, DCopt) are introduced in detail. After that, the software of practical constraint principal curve algorithm is developed, and three algorithms are verified and analyzed by using simulated data and measured GPS data.

The results show that the improved practical constrained principal curve algorithm performs well in calculation accuracy, calculation efficiency, data storage, complex track fitness, and algorithm stability.

GPS digital map made by using the practical constraint principal curve algorithm has a good application prospect. It can help train positioning in train control and reduce data storage space at the same time. Therefore, it is also the future research direction to make GPS digital map by using the practical constrained principal curve algorithm.

References

1. Zhang J, Wang J. An overview of principal curves. Chin J Comput. 2003;26(2):129–677.
2. Hastie T. Principal curves and surfaces. Stanford Univ CA Lab for Computational Statistics; 1984.
3. Hastie T, Stuetzle W. Principal curves. J Am Stat Assoc. 1989;84(406):502–16.
4. Banfield JD, Raftery AE. Ice floe identification in satellite images using mathematical morphology and clustering about principal curves. J Am Stat Assoc. 1992;87(417):7–16.
5. Kegl B. Principal curves: learning, design, and applications. Ph.D. dissertation. Concordia University, Canada; 1999.
6. Kégl B, Krzyzak A, Linder T, et al. A polygonal line algorithm for constructing principal curves. In: Proceedings of advances in neural information processing systems. Denver Colorado, USA, 1999. pp. 501–07.
7. Kégl B, Krzyzak A, Linder T, et al. Learning and design of principal curves. IEEE Trans Pattern Anal Mach Intell. 2000;22(3):281–97.
8. Reinhard K, Niranjan M. Subspace models for speech transitions using principal curves. Proc Inst Acoust. 1998;20(6):53–60.
9. Reinhard K, Niranjan M. Parametric subspace modeling of speech transitions. Speech Commun. 1999;27(1):19–42.

10. Hermann T, Meinicke P, Ritter H. Principle curve sonification. In: Proceedings of international conference on auditory display, Atlanta, USA, 2000. pp. 81–6.
11. De'ath G. Principal curves: a new technique for indirect and direct gradient analysis. *Ecology*. 1999;80(7):2237–53.
12. Dong D, McAvoy TJ. Nonlinear principal component analysis-based on principal curves and neural networks. Comput Chem Eng. 1996;20(1):65–78.
13. Wilson DJH, Irwin GW, Lightbody G. RBF principal manifolds for process monitoring. IEEE Trans Neural Netw. 1999;10(6):1424–34.
14. Seung HS, Lee DD. The manifold ways of perception. Science. 2000;290(5500):2268–9.
15. McFarland JL, Fuchs AF. Discharge patterns in nucleus prepositus hypoglossi and adjacent medial vestibular nucleus during horizontal eye movement in behaving macaques. J Neurophysiol. 1992;68(1):319–32.
16. Taube JS. Head direction cells and the neurophysiological basis for a sense of direction. Prog Neurobiol. 1998;55(3):225–56.
17. Duchamp T, Stuetzle W. Extremal properties of principal curves in the plane. The Annals of Stat. 1996;24(4):1511–20.
18. Duchamp T, Stuetzle W. Geometric properties of principal curves in the plane. In: Robust statistics, data analysis, and computer intensive methods. New York, NY: Springer; 1996. pp. 135–52.
19. Verbeek JJ, Vlassis N, Kröse B. A k-segments algorithm for finding principal curves. Pattern Recogn Lett. 2002;23(8):1009–17.
20. Zhang J, Chen D, Kruger U. Adaptive constraint K-segment principal curves for intelligent transportation systems. IEEE Trans Intell Transp Syst. 2008;9(4):666–77.
21. Chen D, Cai B, Tang T. An information fusion algorithm for multiple GPS track data. In: Fourth international conference on natural computation, IEEE, vol. 6; 2008. pp. 468–72.

Chapter 8
Object-Oriented Digital Track Map of Train Control System Based on Graph Theory

8.1 Overview

8.1.1 GPS Digital Map

With the modernization of transportation technology, the original paper maps and train positioning methods are more and more difficult to meet the technical requirements. Electronic map and navigation, computer graphics, database, geographic information system, and other advanced technology combine the geographical position and the relevant properties. According to the actual needs, the positioning information is output to the user accurately, truthfully, and visually, so as to meet the user's requirements for spatial information. In recent years, the combination of the accuracy of GPS navigation positioning with the directness of electronic map and the rapid processing of computer has brought navigation and control technology into a new field and been applied in many fields [1], which has become a development direction of train control. At the same time, more and more GNSS services have a significant impact on train positioning. In rail transit, the use of GNSS technology for train positioning can reduce trackside equipment and achieve interoperability and safety at a very low cost [2].

At present, in order to improve the efficiency of transportation, the digital track map technology applied in railway transportation system is studied and applied at home and abroad. The reasonable digital track map structure is the basis of the advanced technology of train control system, such as GPS, navigation positioning, and high-speed computer processing. In addition, the digital track map is applied to the onboard equipment and the train control center to facilitate the understanding of various information of the route. Digital track map applied to train positioning can effectively correct errors and improve reliability. Keeping balise information in digital track map can realize virtual balise and greatly reduce the cost. The information of the signal machine is stored in the digital track map, which can realize the virtual signal machine and reduce the ground equipment [3].

D. Chen and R. Cheng, *Intelligent Processing Algorithms and Applications for GPS Positioning Data of Qinghai-Tibet Railway*, https://doi.org/10.1007/978-3-662-58970-0_8

In this chapter, by combining electronic map and geographic information system, the design and realization of digital track map are studied according to specific requirements in the field of rail transit.

With the development of GIS technology, the advantage of applying this technology to rail transit is more and more obvious. Train positioning and control system using electronic map technology can reduce trackside equipment, reduce operation cost, and maintenance cost, and improve the reliability and real-time of positioning system. At present, some European countries and the USA have conducted a series of studies on the application of GPS technology to generate digital track map database and proposed different track map design methods and database structure according to different situations in various regions.

Since May 2003, the West Coast Route Modernization (WCRM) Program in the UK has updated a large number of railway lines. In practice, by combining previous research results, they developed a set of integrated real-time dynamic (Real Time Kinematic, RTK) GPS multi-sensor measurement system—Swiss carts. The cart is manually pushed to obtain data during the surveying and mapping, and then the obtained data is processed to finish gross error detection, data reduction, filtering and smoothing, etc. The problem of this method is that it requires human participation; high demand and high cost for GPS equipment positioning accuracy; the measurement times are limited [4].

The LOCOPROL project in the USA adopts the satellite-based one-dimensional train positioning algorithm [5] to study the generation method and specification of track map, which is composed of object data and track data. Among them, the objects include grade crossing, speed limit sign, and other related points. The track data is the three-dimensional coordinate data information of the railway line and the organization data under the special topological structure and geometric model. Route surveying and mapping uses surveying and mapping equipment installed on the train to carry out GPS surveying and mapping. In addition, the track is surveyed and mapped in the process of train driving, and then the digital track map is synthesized by calculating in sections.

The DBGIS database used by German railways is applied to railway construction and line maintenance, but it is not optimized according to the demand of train positioning, and the data storage is dispersed, which is not conducive to the use of train control system. Open Track project of ETH Zurich studies an object-oriented map database that is suitable for train simulation. The content of this database far exceeds the basic content such as detailed line geographic information, and it adopts the non-universal WGS-84 coordinate system [6].

The ITCS train control system adopted by Qinghai–Tibet railway includes the onboard map database, which contains the line description data, train operation data, communication data, and other contents. The more detailed information includes the start and end position of the whole line, number of tracks, ITCS operational zones' position, operational train data, communication network data, and so on. By using the data saved in the database, the virtual signal machine is realized, ground equipment is reduced, and the station is unattended [7].

In China, the research on digital track map started relatively late, and there was even no digital track map in the railway construction in the early stage. So, there was still a lack of research on digital track map in China. But similar studies have been done on navigation maps in other transportation systems. For example, in urban traffic, navigation map data is generally divided into two categories according to its nature: graphic data and topology network data. There are two types of data in the topology network: road type (line) and information point type (point), which constitute the path data of the abstract urban traffic network. According to the hierarchical structure similar to tree in graph theory, path information stores the three-level map of urban road network, provincial road network, and national road network. When querying, locate the section by the order from top to bottom [8].

Digital navigation map in electronic map is also a kind of vector map. Generally, a search index tree (such as R-tree or quadtree) should be established according to a certain index algorithm to indicate the location of each ground object. Then, the index tree should be searched according to the geographical range to read the longitude and latitude information contained therein. By using coordinate transformation algorithm (such as Gaussian Kruger algorithm), the geographic longitude and latitude coordinates were converted to Gaussian plane coordinates to display the aircraft on the electronic map [9].

For rail transit, there have been studies on digital track maps. Among them, the polyline representation method of linear orbit and curved orbit is discussed for the long-distance track, and the generation method of electronic map is proposed by using both accurate GPS data and real-time original data [10]. But the division of track segment is artificial and the error is large. In addition, previous papers focus on the preprocessing of data and the format definition of data, which is only suitable for train positioning and display. However, the electronic map is less combined with train control.

The train control system map is composed of basic element nodes and zones by applying topological mathematical model, which defines different ways of track connection. When the node connects up to three zones, the topology connection properties of the node for the upstream and downstream zones of the node are defined. When the real intersection connects more than four (including four) sections at the same time, the auxiliary virtual node is defined to complete the connection [11].

8.2 Design of Digital Track Map

When designing the digital track map, the first problem to be solved is the GPS data processing. The generation process of digital track map includes five steps: route mapping, data processing, model selection of map data, compilation, and data verification. Although the data processing contains many contents, the GPS measurement itself has drift error. Moreover, in order to reduce the cost of measurement, the precision of the used GPS instrument is not high, which leads to the data precision is not up to the requirement of making digital track map. At this time, some algorithm is needed to preprocess the data to remove the data with large error. When making digital track map, the fitting of railway track data is a very important problem, which affects a series of problems such as the precision, storage mode, memory occupation, and positioning algorithm of digital track map.

Error processing and data fitting are a complex process, as the topic of this chapter is completed under the funding of Chinese National Natural Fund project "key algorithms and methods for automatic generation of train control digital track map." Error processing and data fitting are completed by others group in the project [12, 13], so this chapter is a study based on the assumption that the data has been processed. Curve linearization is chosen as the method of curve fitting in this project, that is, for some non-linear sections, multiple straight approximation curves that meet certain error requirements can be selected. Therefore, the digital track map only needs to store the information of line segment, which reduces the complexity of calculation. In this chapter, each line segment obtained by fitting method is taken as the smallest unit of computation, in this case it is called "the segment."

8.3 Overall Design

Since both the topological data model and the object-oriented data model have advantages and disadvantages (analysis is shown in Table 8.1), if only one of them is adopted, the complex actual geographical environment cannot be described, or the expandability and the maintainability of the designed data model is not good, and the data redundancy is large. Meanwhile, in the theory of graph theory, many practical geographic problems can be abstracted into the geographical network, that is, the arrangement pattern of points and lines. The basic definition of a graph consists of two basic elements: the point set (or vertex set) and the edge set (or arc set). By combining the point set and edge set of graph theory with both the object-oriented data model and the topological relation data model, we cannot only take advantage of their advantages, but also avoid their disadvantages. Therefore, in the rail transit, the track sections and their mutual connections are described as the geographic network in the sense of graph theory through some simplification and abstraction process. The geographical location of the signal machine, turnout, and segmentation points of track sections can be abstracted into the points in the graph theory and designed with the object data model. The tracks and their interrelations are abstracted into the links of points and points, that is, edge sets, which can be designed by topological relational data model.

Based on the above analysis, the designed digital track map has two the database: One is a spatial database, which mainly stores the spatial organization information of digital map; the other is the object property database, which contains the information of the special turnouts, signal machines, speed limit marks, and kilometer marks in this system. The design of the spatial database and object property database is described in detail below. Figure 8.1 shows the whole generation process of digital track map.

Table 8.1 Comparison of two data models

	Object-oriented data model	Topology data model
Independence	Strong	Weak
Efficiency for individual entities	High	Low
Partial update	Easy	Difficult
Topology relationship	Loose, temporary building	Clear, compact
Network analysis efficiency	Low	High
Common points and common sides	Repeat storage, data redundancy	No data redundancy
Expressive complex entity	Easy	Difficult

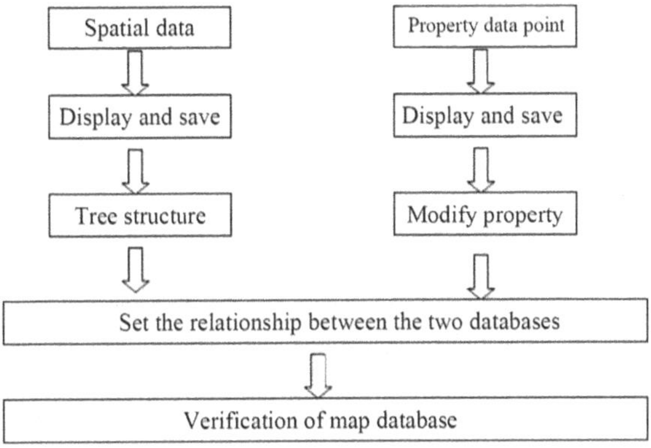

Fig. 8.1 Digital track map generation process

8.3.1 Spatial Database Design

The design of spatial database includes the design of logical structure and storage structure. Logical structure includes linear structure, tree structure, and graph structure. A tree is a hierarchy defined by a branching relationship, and a tree structure is a structure where nodes not only have branches but also have hierarchical relationships. In rail transit, tree structure is similar to station structure, and this data structure can be applied to digital track map. Digital track map adopts a tree structure, which has clear hierarchy and can organize station equipment in an orderly way, making the structure simple, clear and convenient for computer storage. The structure of the whole station is defined by the nature and characteristics of the tree in the data structure, and each group of data is numbered and clearly dependent on each other. As shown in Fig. 8.2, the station is divided into four layers: The first layer is the railway network, such as the national network and a regional network. The second layer is the station and operational zone. The third layer is the spatial data and object attribute data information included in the station and operational zone. The fourth layer is the storage data of the digital track map. The object property data includes the data of the important equipment in the station.

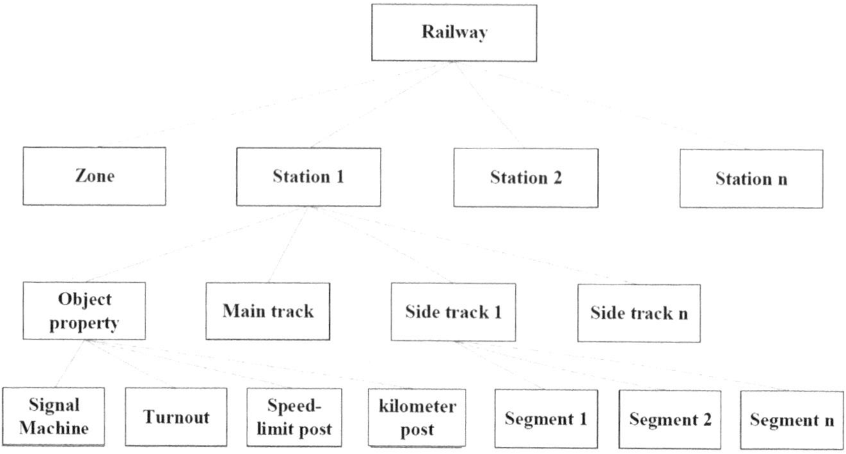

Fig. 8.2 Station logic diagram

Storage structure includes sequential storage structure and chain storage structure [9]. In the design process of the logical structure of the tree structure, as the number of subtrees contained in each layer of the railway organization structure is not necessarily fixed, the linked list structure is selected as the storage structure of the tree in this paper, and each node in the linked list contains two pointer fields and one data field. For the logical structure, the nodes in the first three layers usually appear as root nodes, and their data domains can store the number of lower nodes and the scope of railway and station, so as to facilitate later search. The third layer is a collection of spatial data and attribute data. The spatial data of layer 4 mainly includes railway line information. The tree structure is shown in Table 8.2.

Table 8.2 Tree structure description

Level number	Data domain	Pointer field 1	Pointer field 2
First level	Information	–	–
Second level	Information	Point to the head point of the next layer	Point to the next point of the same layer
Third level	Information	Point to the head point of the next layer	Point to the next point of the same layer
Fourth level	Segment data	The in-line connection points to the next node of the same root	The in-line connection points to the next node of the different root

8.3.2 Object Property Database

Based on the overall design of the digital track map described above, there are some important devices in rail transit that need to be designed independently. Object-oriented data model is designed by using the object-orientation method. In this chapter, the design of signal machine, turnout, kilometer mark, and speed limit mark are mainly considered.

There are many kinds of signal signals according to the usage. The most basic signal includes shunting signal, arrival signal, route signal, and starting signal. The rest are hump signal, warning signal, protective signal, blocking signal, etc. [10]. In this chapter, we consider only three basic signal types. Table 8.3 shows the static attribute data of the signal. Similarly, turnout, kilometer mark, and speed limit mark can be designed based on Tables 8.4, 8.5, and 8.6.

Table 8.3 Signal properties' representation

Label	ID	Sg_point	Light	Height	Forbid	Direction
Meaning	Signal machine ID	Coordinates of point	Type	Height type	Forbidden light	Display direction
Type	Int	CPoint class	Bool	Bool	Bool	Bool

Table 8.4 The composition structure of turnout

Label	ID	Sw_point	Flag
Meaning	Turnout ID	Coordinates of point	Turnout state
Type	Int	CPoint class	Bool

Table 8.5 Kilometer mark structure

Label	ID	ks_point	Flag
Meaning	Speed limit post ID	Coordinates of point	Mileage of coordinate point
Type	Int	CPoint class	Double

Table 8.6 Speed limit mark structure

Label	ID	ks_point	Flag	Information
Meaning	Speed limit post ID	Coordinates of point	Mark whether the speed-limit starts or ends	Mileage of coordinate point
Type	Int	CPoint class	Bool (true: speed limit starts, false: speed limit ends)	Double

8.3.3 Spatial Data and Object Attribute Data Match

The main function of railway signal is to ensure the safety of operation. Within railway station, the safety of the train and the route safety should be ensured. In a word, the following measures must be taken to ensure the safety of the train: The route is free before the train enters the corresponding zone; all the turnouts on the route are in the correct position and locked in the correct position to prevent derailment of a moving train by shaking or moving the turnout; it must be ensured that other trains do not crash into the corresponding route from the front, side, and rear direction. Only if the above three conditions are met simultaneously, the control signal can be open.

Thus, interlocking is defined as "by using technical approach, signal machines, turnouts, and routes are controlled to follow certain procedures and meet certain conditions before they can act or establish a mutual relationship." The definition contains two meanings. On the one hand, in order to ensure the safety of station and train driving, a series of interlocking rules must be established to restrict the movement and establishment of signal machines, turnouts, and routes. On the other hand, the rules must be implemented by technical means [28]. Although the rules on interlocking are not discussed in this chapter, it is also necessary to combine the object attributes of signal machines, turnouts, and so on that affect the train's movement with both the spatial data mentioned in the previous chapter and the train's movement position. This design method has two functions: One is to ensure the safety of the train, and the other is to be able to real-time query the state of the train or the signal machines and the turnouts in front zones of running train.

Based on the above analysis, in order to ensure traffic safety, spatial data and attribute data must have a certain connection. In addition, certain sign markers must be set in the structure of spatial data to realize the connection between spatial data and attribute data.

In the design of spatial data, the composition of segments is mentioned. Segments are composed of line segments between two points and the data information of data points is needed in the structure of fragments (see Table 8.7). For the attributes of the data point, two attribute values are added to set the connection between the spatial data and the attribute data. The connection between spatial data and attribute data consists of two steps. The first step is to add a turnout marker and a signal marker in the storage structure of each data point. The added markers represent the object attributes that match the signal machines, turnouts, and so on.

The composition structure of the data node is shown in Table 8.7. The second step is to design the matching algorithm. Based on the feature that a certain error value is met between the object attributes and data points, a flow chart of the matching algorithm is designed and shown in Fig. 8.3.

Table 8.7 Composition structure of data nodes

Label	Type	Meaning
ID	Int	ID number of the point
Point class	CPoint	Coordinates of memory points
Signal_flag	Bool	Ture: the point is set up with signal machine; False: no signal machine
Signal_Up	Int	Uplink signal machine's signal
Signal_Down	Int	Downlink signal machine's signal
Switch	Int	The number of the turnout

8.4 The Digital Track Map System and Its Verification

The digital track map system mainly includes four modules: map display module, spatial data organization module, object attribute data organization module, and map verification module. The composition modules of the track map system are shown in Fig. 8.4.

The purpose of digital track map verification is to check whether the connection relation of spatial database in digital track map is correct, check whether the setting of the object property database in the digital track map is reasonable, verify the realization of mutual query between spatial database and object attribute database in digital track map. The main flow diagram of the validation algorithm is shown in Fig. 8.5. Based on the idea of depth-first search and breadth-first search in graph theory, the verification algorithm of digital track electronic map is designed. The basic steps are described as follows: given the position of the initial node and the destination node, a route is selected and the state of the relevant object attributes should be queried to determine whether the digital track map is connected. During the execution of the algorithm, the first step is to initialize the search content, including setting the search marker of all line segments to false before searching. The search marker is set up to prevent an endless loop and indicates that all line segments have not been searched. In addition, it is necessary to check whether the conditions of starting line segment are met. If the conditions are met, enter the data into the stack and record the ID number of line segment. If not, program stops and search process ends. Then, the program starts from the input starting line segment and visits the adjacent line segment that has not been accessed by determining whether the adjacent line segment is the line segment of the endpoint. If it is not the adjacent line segment, the program accesses the adjacent line segment and repeats the process until the line segment of the endpoint is accessed. It also completes the query of the object property database.

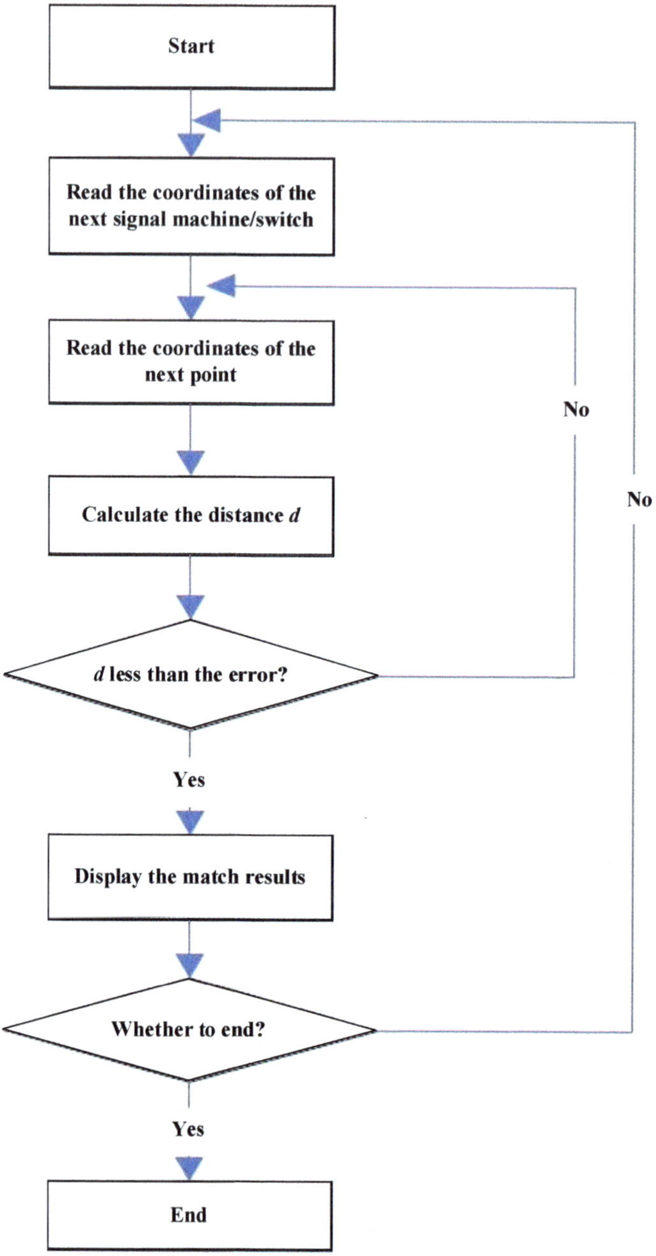

Fig. 8.3 The flow chart of matching algorithm of property data point

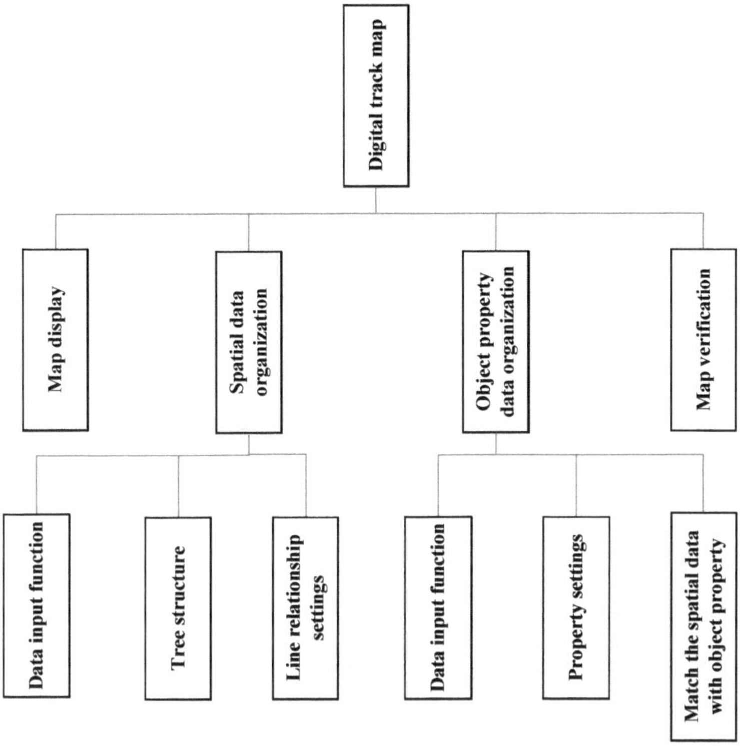

Fig. 8.4 The composition modules of the track map system

8.5 Application of Satellite Positioning Technology

Over the last decade, important efforts have been made with regard to satellite-based positioning systems in order to adapt them to civilian applications (e.g., pedestrian and vehicle navigation), such as the development of Galileo (Europe) and BeiDou (China). The last three years of European research projects that have been focused on train positioning are presented in the following parts [14].

The railway sector has always been present in different European research programs developed in recent years (e.g., FP7 and Horizon 2020). Horizon 2020 is the biggest European research and innovation program ever [15]. It is divided into three main areas, as shown in Fig. 8.6. Within it, all the investigation activities related to the railway are collected in Shift2Rail (S2R) [16]. S2R is the first European rail initiative that seeks research and innovation market-driven solutions to be integrated into innovative rail products [16, 17]. In order to tackle more specific objectives, five innovation programs (IP) have been designed, as shown in Figs. 8.6 and 8.7, where train navigation is tackled in the IP2 (Advanced Traffic Management and Control Systems).

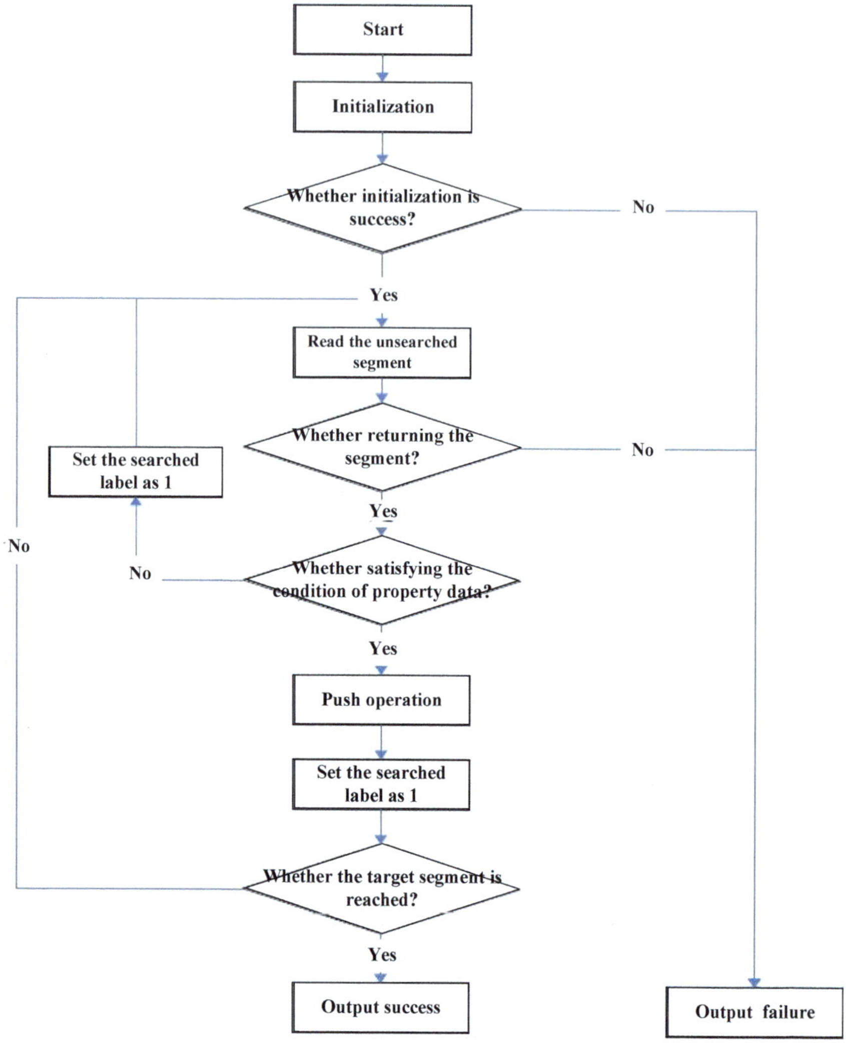

Fig. 8.5 The main flow diagram of the validation algorithm

ERSAT-EAV (ERTMS on Satellite Enabling Application and Validation) is the very first project launched inside S2R, as shown in Fig. 8.7. The purpose is to verify the suitability of GNSS as the enabler of a cost-efficient and economically sustainable ERTMS signaling solution for railway safety applications [18]. It includes a measurement project, but it goes further, looking for a systematic solution (implemented, tested, and validated for safety applications).

RHINOS (Railway High Integrity Navigation Overlay System) is focused on applying GNSS in the railway context while ensuring architectures of high integrity.

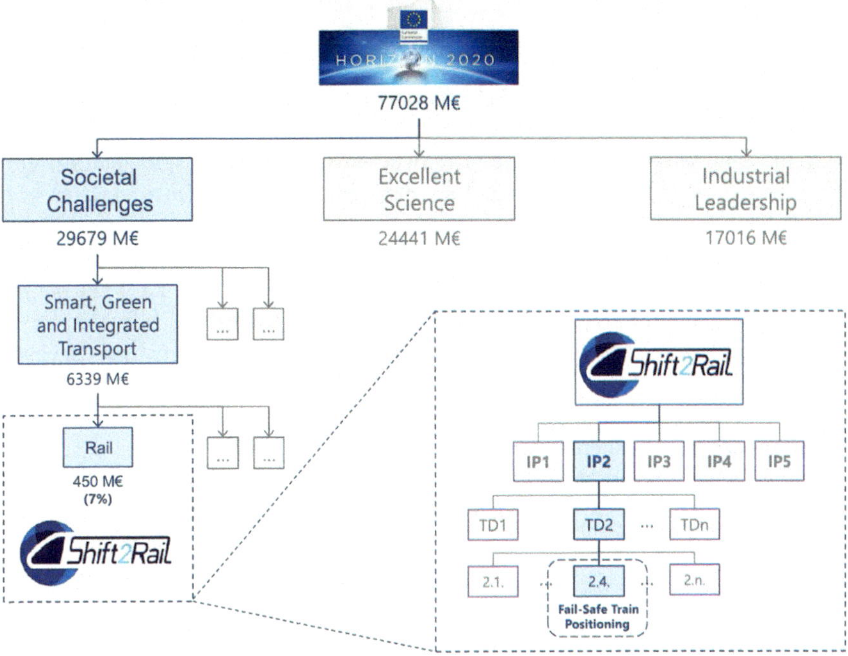

Fig. 8.6 Horizon 2020 and S2R breakdown

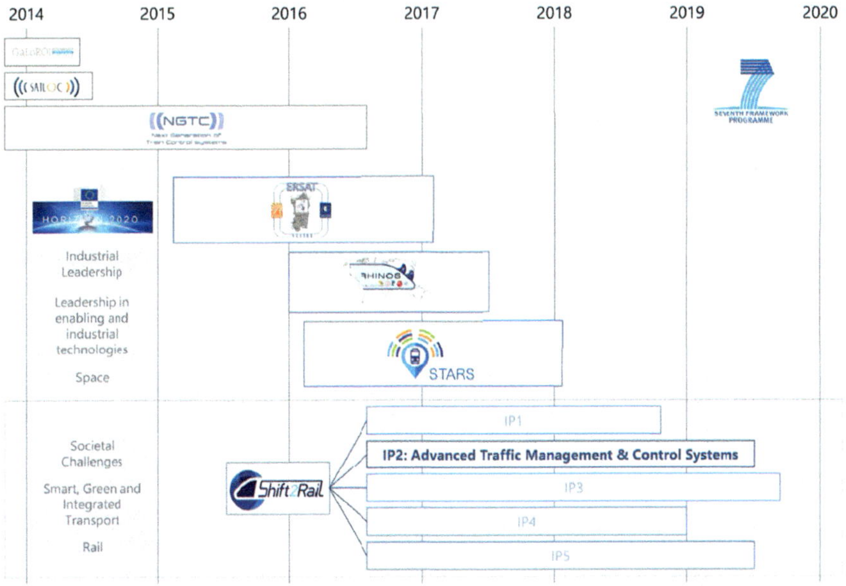

Fig. 8.7 Chronological train positioning context inside S2R and previous projects (GaLoROI, SATLOC, NGTC, ERSAT, RHINOS, and STARS) feed

Its objective is to increase the use of GNSS to support the safety-critical train localization function for train control in emerging regional and global markets [19].

STARS (Satellite Technology for Advanced Railway Signalling) is meant to fill the gap between the needs of an ERTMS for safety-critical applications and GNSS services. The key objectives are the development of a universal approach to predict the achievable GNSS performance in railway environments and the quantification of the economic benefits this technology can bring [20]. In essence, the aim is to include GNSS into ERTMS, maintaining both the safety and the interoperability of the system. The input of this project is the different work done in NGTC and ERSAT projects as well as older ones (GRAIL, LOCOPROL, LOCOLOC, etc.). The outputs should be the analyzed data obtained from different railway tests. A further survey of the research projects about the inclusion of GNSS in railway signaling systems is collected in [21].

GNSS presents limitations in signal-denied areas, such as tunnels, urban environments, and foliage. In such areas, no GNSS solution is available. Consequently, the safety integrity of the positioning system is jeopardized. The limitations of the availability of a GNSS signal in those areas are one of the main reasons why extra sensors are needed. Thus, researchers have focused on its integration via multisensory data fusion.

The very first article on data fusion related to train navigation presents candidate sensors for positioning and outlines a generic sensor fusion architecture [22]. The algorithms and sensors presented set the course for any later research, although some of the characteristics and technologies presented are nowadays obsolete (e.g., the selective availability in GPS signal and best accuracy reached, the nonexistence of an IMU technology based on Microelectromechanical systems (MEMS), a maximum sampling frequency of GNSS limited by 1 Hz).

Since that first publication, and in parallel with the development of technology (e.g., new GNSS constellations, the inclusion of MEMS in consumer electronics, etc.), many research articles have been published. Many authors have focused on the localization of land vehicles and have extrapolated the architectures designed from this area to the railway context. Most of the literature available up to date are based on this idea [23–25]. Although trains are also land vehicles, they have special features (railway track constraints, slip and slide phenomena, smooth changes in trajectories and altitude, etc.) that necessitate a special architecture. For example, a GPS and IMU fusion architecture is proposed in [26]. It presents an explicit explanation of the equations needed to model the system as well as the results of field tests. It can be concluded that the same architecture nowadays will give more accurate results since the selective availability of GPS signal is now disabled.

Other authors have tackled the train positioning problem by taking into account the dynamics of the train and the restrictions imposed by the track. The most relevant work in this sense has been done in a collaboration between the University of Florence and the University of Siena [27]. An implementation of a tachometer and a 6-DOF IMU was carried out combined with a train multibody model and the computation of multiple Kalman filters. The solution had a maximum error of about 20 m in travelled distance and 1.5 m/s in calculated speed, so it fulfills the European Train Control

System (ETCS) requirements in terms of accuracy according to [28]. This work has some preceding papers, where the state of the art of railway sensors was presented and a Tachometer-IMU architecture discussed, building a railway vehicle dynamic model and developing a system with improved wheel velocity measurements [29–31].

A rather original standpoint is presented in [32–35]. Instead of 3D positioning, the authors consider a train as a unidimensional vehicle constrained by a track. Thus, the algorithm only needs to solve a 1D problem. Thanks to track map information, the one-dimensional solution is then linked to a 3D map and the train can be located in absolute coordinates.

8.6 Brief Summary

GIS technology is a new emerging technology in recent years. It combines a series of technologies such as electronic atlas and navigation, computer graphics, database and geographic information system, and organically combines geographic location and related attributes. At the same time, according to practical requirements, the map and the information contained in the map are output to the user in real time and accurately. Because of the above advantages of GIS technology, it has certain application value in rail transit. In recent years, the rapidity of GPS navigation and positioning, the directness of GIS electronic map, and the fast processing of computer have been combined to open up a new development direction for railway positioning and navigation technology.

This chapter combines the basic model of graph theory, the basic idea of object orientation and the characteristics of rail transit to carry out the research, and realizes the model design, realization and verification of digital track map, which is applicable to the practical application of railway train control system. The chapter mainly completed the following work:

(1) By using the geographic network topology model, the digital track map is modeled through comparison and selection algorithm. This chapter divides the design of digital orbit map into two parts: the spatial data part and the object attribute data part. At the same time, the spatial data and object attribute data of digital track map are analyzed by combining with the characteristics of rail transit. Then, the models of spatial data and object attribute data as well as their mutual matching and query operations are designed are realized. By using the depth-first search algorithm in graph theory and combining with the object attribute features in train control, a digital track map verification algorithm is proposed.
(2) According to the digital track map model designed in this chapter, the functional classification of digital track map is analyzed, and the system contains four main parts: spatial data structure, map display, object property setting, and map verification.
(3) The verification algorithm is designed to check the digital track map.

As mentioned above, we mainly study the data model of digital orbit map. However, there are still some imperfections in the study. Specifically, there are mainly the following aspects:

(1) optimize human-computer interaction interface to make human–computer interaction process more reasonable;
(2) at present, in order to simplify the research, the verification algorithm of digital track map only considers the problem of one-way search. The next step is to optimize the search algorithm so that the graph search can complete two-way search;
(3) improve the data management mode, realize the data optimization management by using the database, and lay the data foundation for the actual demand.

References

1. Bishop R. A survey of intelligent vehicle application. In: Proceedings of IEEE intelligent vehicles symposium; 2000.
2. Barbu G. Requirements of rail applications. GNSS Rail User Forum; 2000, p. 5.
3. Bedrich S, Gu X. GNSS-based sensor fusion for safety-critical applications in rail traffic. Galileo and EGNOS Information Catalogue; 2004, p. 8.
4. Glaus R, Peels G, Muller U, et al. Precise rail track surveying. GPS World. 2004, 15(5).
5. Track Data Base Generation, [EB/OL], http://www.honeywell.com/sites/Honeywell/1993.
6. Gomm H. Data format specification for the track database. LOCOPROL; 2002.
7. Li K. The project of ITCS signal control system of Qinghai–Tibet Railway. Chinese Railway. 2005;7:31–6.
8. Chen Y, Wang G, Chen B. Research on GPS positioning data in map matching of vehicle navigation. Ind Control Comput. 2006;19(1):61–2.
9. Xiong J, Cheng W. Implementation of GPS receiving and E-map with VC and MapX component. J Air Force Radar Acad. 2006;20(1):51–4.
10. Gao G, Cai B. Research on the automatic electronic map generation algorithm for the train supervision system. J China Railw Soc. 2006;1:63–70.
11. Wang J, Cai B, Wu J. Railway database for train control system. ITS Commun. 2005;7(4):20–2.
12. Chen D, Cai B, Wang J, Tang T. Mathematical model and algorithm of data reduction for railway GPS data. J China Railway Soc. 2008;30(4):116–9.
13. Cheng J. Design and realization of railway signal equipment inspection management information system. Beijing: Beijing Jiaotong University; 2007.
14. Otegui J, Bahillo A, Lopetegi I, et al. A survey of train positioning solutions. IEEE Sens J. 2017;17(20):6788–97.
15. Horizon. [Online]. 2020. Available: https://ec.europa.eu/programmes/horizon2020/en/background-material.
16. Shift 2 Rail. [Online]. Available: http://shift2rail.org/.
17. Economou C. Shift2rail JU update and the role of the lighthouse projects, Shift2Rail, Brussels, Belgium. Technical Report, May 2015.
18. ERSAT EAV, Inno Trans, Berlin, Germany, 2016.
19. RHINOS. Accessed: 7 Sept 2017. [Online]. Available: https://www.gsa.europa.eu/railway-high-integrity-navigation-overlay-system.
20. STARS. Accessed: 7 Sep 2017. [Online]. Available: http://www.starsrail.eu/project/.
21. Marais J, Beugin J, Berbineau M. A survey of GNSS-based research and developments for the European railway signaling. IEEE Trans Intell Transp Syst (to be published). [Online]. Available: http://ieeexplore.ieee.org/document/7857080/.

22. Mirabadi A, Mort N, Schmid F. Application of sensor fusion to railway systems. In: Proceedings of 1996 IEEE/SICE/RSJ international conference multisensor fusion integration intelligent systems, Dec 1996, pp. 185–192. [Online]. Available: http://ieeexplore.ieee.org/abstract/document/572176/.
23. Wei Z, Ma S, Hua Z, Jia H, Zhao Z. Train integrated positioning method based on GPS/INS/RFID. In Proceedings of 2016 35th IEEE Chinese control conference (CCC), pp. 5858–5862. [Online]. Available: http://ieeexplore.ieee.org/abstract/document/7554274/.
24. Reimer C, Müuller FJ, Hinüber ELV. INS/GNSS/odometer data fusion in railway applications (ISS), Karlsruhe, Germany, vol. 2; 2016, pp. 1–14.
25. Veillard D, Mailly F, Fraisse P. EKF-based state estimation for train localization. In: Proceedings of 2016 IEEE SENSORS, Oct/Nov 2016, pp. 1–3. [Online]. Available: http://ieeexplore.ieee.org/abstract/document/7808726/.
26. Cai B, Wang X, editors, Train positioning via integration and fusion of GPS and inertial sensors. In: WIT transactions on the built environment. Southampton: WIT; 2000.
27. Malvezzi M, Vettori G, Allotta B, Pugi L, Ridolfi A, Rindi A. A localization algorithm for railway vehicles based on sensor fusion between tachometers and inertial measurement units. Proc Inst Mech Eng F J Rail Rapid Trans. 2014;28(4):431–48.
28. Performance Requirements for Interoperability (Subset-041), UNISIG, Brussels; 2012.
29. Allotta B, Pugi L, Ridolfi A, Malvezzi M, Vettori G, Rindi A. Evaluation of odometry algorithm performances using a railway vehicle dynamic model. Vehicle Syst Dyn. 2012;50(5):699–724. [Online]. Available: http://www.tandfonline.com/doi/abs/10.1080/00423114.2011.628681.
30. Allotta B, Colla V, Malvezzi M. Train position and speed estimation using wheel velocity measurements. Proc Inst Mech Eng F J Rail Rapid Trans. 2002;216(3):207–25. [Online]. Available: http://pif.sagepub.com/lookup/doi/10.1243/095440902760213639.
31. Malvezzi M. Odometry algorithms for railway applications. Ph.D. dissertation. Bologna: Department of Applied Mechanics, University of Bologna; 2003.
32. Heirich O, Robertson P, Garcia AC, Strang T. Bayesian train localization method extended by 3D geometric railway track observations from inertial sensors. In: Proceedings of 15th international conference on information fusion (FUSION), July 2012, pp. 416–423. [Online]. Available: http://ieeexplore.ieee.org/abstract/document/6289833/.
33. Heirich O, Lehner A, Robertson P, Strang T. Measurement and analysis of train motion and railway track characteristics with inertial sensors. In: Proceedings of 14th international IEEE conference on intelligent transportation systems (ITSC), Oct 2011, pp. 1995–2000. [Online]. Available: http://ieeexplore.ieee.org/abstract/document/6082908/.
34. Crespillo OG, Heirich O, Lehner A. Bayesian GNSS/IMU tight integration for precise railway navigation on track map. In: Proceedings of IEEE/ION position, location and navigation symposium (PLANS), May 2014, pp. 999–1007. [Online]. Available: http://ieeexplore.ieee.org/abstract/document/6851465/.
35. Heirich O. Bayesian train localization with particle filter, loosely coupled GNSS, IMU, and a track map. J Sens. 12, 2016, Art. no. 2672640. [Online]. Available: http://www.hindawi.com/journals/js/2016/2672640/.

Chapter 9
Summary and Future Research

9.1 Summary

The completion and opening of the Qinghai–Tibet railway shorten the space and time gap between the Qinghai–Tibet region and other parts of the country and promote the integration of this region into the national market. The change of transportation and information conditions will directly drive the development of local agriculture and animal husbandry and improve the ability of relevant industries to resist risks and disasters. After the Qinghai–Tibet railway is put into operation, it is expected that 75% of the freight volume will be borne by the railway in the future. According to the designed transportation capacity of the Qinghai–Tibet railway, the one-way transportation capacity of the Qinghai–Tibet railway is more than 40 times of the current automobile freight transportation capacity of the whole region of Tibet. The opening of the Qinghai–Tibet railway will reduce the transportation cost by more than half, thus thoroughly breaking the bottleneck of Qinghai–Tibet transportation and greatly alleviating the situation of tight transport capacity.

The Qinghai–Tibet railway links Tibet more closely with other parts of the motherland by convenient, rapid, and economic means of transportation. The journey from Beijingwest to Lhasa takes only 47 h to arrive, and the ticket price for hard seats is 389 yuan. People from the ministry of commerce predict that as more and more colorful, high-quality, and cheap goods enter Tibet via the Qinghai–Tibet railway, and the purchasing power of Tibetan people will be greatly enhanced. Zhou Tianyong, a professor at the party school of the central committee of the Chinese Communist Party, said that with the opening of the Qinghai–Tibet railway and the strengthening of foreign exchanges, more and more Tibetans will go out to study, visit, do business, and work and have closer ties with other provinces and regions. Experts from the Ministry of Construction of China believe that the opening of Qinghai–Tibet railway is a rare opportunity for the development and construction of small towns in Qinghai–Tibet region. Like a golden ribbon, the Qinghai–Tibet railway connects the cities and towns with rich resources and distinctive features. According to statistics,

D. Chen and R. Cheng, *Intelligent Processing Algorithms and Applications for GPS Positioning Data of Qinghai-Tibet Railway*, https://doi.org/10.1007/978-3-662-58970-0_9

after the opening of the Qinghai–Tibet railway, the passenger flow from Xining alone will reach more than 600,000 per year. The people and logistics will make the pearl of this "golden belt" shine brightly. The small towns along the belt will fully show the "world's third pole," which reflects the overall appearance of the Qinghai–Tibet plateau from one to another, and the towns along the Qinghai–Tibet line will become increasingly prosperous.

The newly opened Qinghai–Tibet railway, with an investment of 1.54 billion yuan for environmental protection, has become an unprecedented "mega-green project" in the history of China railway construction. Moreover, the completion of the Qinghai–Tibet railway can transport rich coal, oil, and other resources in northwest China into Tibetan areas through economic and convenient railway channels, so as to meet the energy demand of people in agricultural and pastoral areas. This is of great significance for curbing acts of destroying forests and grasslands and protecting the ecological environment. After the Qinghai–Tibet railway was put into operation, it has become the main mode of tourism and material transportation in and out of Tibet, so as to divert the huge transportation fleet of Qinghai–Tibet highway, reduce automobile exhaust emissions, and further improve the atmospheric quality of the Qinghai–Tibet plateau.

As a key technology for the operation of Qinghai–Tibet railway, GPS satellite positioning is the basis for ensuring the safety, efficiency, and smooth operation of Qinghai–Tibet railway trains. Therefore, the processing of collected GPS data is the direction of future efforts of scholars. This book addresses the key issues in GPS satellite positioning of Qinghai–Tibet railway: intelligent processing of satellite positioning data, including error data detection, data reduction, multi-GPS track information fusion, and graphics-based and object-oriented digital track map. In addition, the authors used a large number of measured GPS data in Qinghai–Tibet railway to carry out research. By using theories and methods such as optimization, expert system, and machine learning, we put forward our own research ideas and relevant results, as a reference, hoping scholars to continuc to put forward more superior methods.

9.2 Future Research

In the current railway satellite positioning, only GPS receiver is used to obtain the navigation information. However, the single sensor to perceive the navigation information (vehicle posture) is difficult to meet the accuracy and reliability required by the autonomous vehicle navigation system. Therefore, we will combine GPS and multi-sensor integrated technology in the future to meet the accuracy and reliability required by autonomous vehicle navigation system.

The independently developed Beidou satellite navigation system in China includes "Beidou No. 1" system and "Beidou No. 2" system. They have the advantages of small number of satellites, small investment, simple and inexpensive user equipment and can meet the needs of China maritime, land and air transport navigation and positioning. In future research, we will combine our theoretical results with Beidou satellite navigation to improve the accuracy and anti-interference of satellite positioning.

Under the condition of high-speed operation of train, the use of satellite positioning method to realize train positioning will lead to the situation that satellite signals cannot be received, which will result in the low positioning accuracy. However, as satellite aided train positioning method, such as train positioning based on query/balise and speedometer, various methods can learn from each other to achieve the best benefit/cost ratio. In future theoretical research, we will also do in-depth research on this situation.

To sum up, at present, only GPS data processing is far from enough. In addition, the optimization at the system level can not only optimize the equipment structure, but also further improve the operation efficiency, so that the train can be operated and dispatched more intelligently, which will bring the whole experience to passengers.

Printed by Printforce, the Netherlands